Navigating Between Truths

Navigating Between Truths

In a World of Competing Perspectives

ALEXANDER STOTER

RESOURCE *Publications* · Eugene, Oregon

NAVIGATING BETWEEN TRUTHS
In a World of Competing Perspectives

Resource Publications
An Imprint of Wipf and Stock Publishers
199 W. 8th Ave., Suite 3
Eugene, OR 97401

www.wipfandstock.com

PAPERBACK ISBN: 979-8-3852-8743-7
HARDCOVER ISBN: 979-8-3852-8744-4
EBOOK ISBN: 979-8-3852-8745-1

Contents

Preface

We live in a world shaped by different ways of understanding reality.

Across philosophy, religion, science, and society, explanations emerge that offer clarity, direction, and meaning. Yet these explanations do not simply coexist. They develop, stabilize, and sometimes come into tension with one another.

This book does not resolve these differences directly. Instead, it takes a step back to examine the patterns that underlie them and how they are formed.

At the same time, new technologies—including artificial intelligence—are transforming how we form meaning. This may reshape the role we play, as well as the methods we use in our search for what we consider to be true.

Across fields, similar structures appear: partial understanding moves toward claims of completeness, meaning becomes stabilized, and complexity is reduced into more defined forms. These processes are not limited to any one domain—they reflect how human understanding develops more broadly.

By focusing on these patterns, it becomes possible to approach questions that extend beyond individual disciplines.

Rather than offering fixed answers, this book explores how philosophy, religion, and science shape the way we understand reality—not as separate domains, but as interacting perspectives that influence how we relate to what we consider true.

It remains grounded in the human experience of searching for meaning, while recognizing that understanding is not fixed, but part of an ongoing process in which insights emerge, shift, and are revised.

The chapters that follow move through different layers of understanding—systems, mind, language, society, education, and the individual—each offering a perspective on how meaning is formed and how it becomes stabilized.

Chapter 1

The Question

We do not begin from certainty, but from a far more immediate and undeniable fact: we find ourselves existing, without fully understanding why. This is not a conclusion reached through reasoning, nor a belief we consciously adopt. It precedes both. It is the simple and unavoidable experience of being present in a reality that we did not create and do not fully comprehend.

In everyday life, this rarely appears as a problem. We move through routines, respond to responsibilities, and orient ourselves toward goals that provide structure and direction. The world presents itself in practical terms: tasks to complete, decisions to make, relationships to maintain. Within this movement, deeper questions tend to remain in the background. They are not necessarily rejected but deferred—set aside in favor of what demands immediate attention.

In contemporary life, this question is often set aside—not because it disappears, but because the demands of daily activity leave little space for it to surface. Yet it remains present, reappearing in moments where routine is interrupted or perspective widens.

And yet, this postponement is never complete.

There are moments—sometimes quiet, sometimes disruptive—when the question returns. It may arise in reflection, in confrontation with loss or limitation, or in a sudden awareness of the scale of the universe. It may emerge not through crisis, but through stillness, when the ordinary flow of activity pauses just long enough for something more fundamental to become visible.

In such moments, the familiar structure of everyday life comes to be less stable, and what had remained implicit begins to surface more explicitly.

The question, when it appears in this way, does not focus on any aspect of reality. It does not ask how things function or how they can be changed. Instead, it addresses something more fundamental: why there is anything at all, and what it means that we are part of it.

This question resists the forms of answer we are accustomed to. It cannot be resolved through practical reasoning, nor does it yield to straightforward explanation. It does not appear as a problem to be solved, but as a condition that remains open. It opens a space rather than closing one. And once that space appears, it emerges as difficult to ignore entirely.

What makes this question particularly striking is that it does not arise from specialized knowledge. It does not require philosophical training or scientific expertise. It emerges from the basic condition of being human. It is accessible yet still exceeds our ability to fully grasp it.

Throughout history, this difficulty has not been overlooked. On the contrary, it has given rise to entire traditions of thought. Philosophy, religion, and science can each be understood, in part, as responses to this underlying problem. Each attempt, in its own way, to clarify the nature of reality and our place within it.

Across traditions of thought, this question has given rise to different ways of approaching reality and our place within it.

Philosophy often seeks to move beyond the instability of appearances toward something more enduring. Thinkers such as Plato [1] proposed that what we encounter through our senses does not exhaust reality, and that true understanding requires access to a more stable and intelligible order. In this view, the uncertainty of experience is not final, but points beyond itself to something more fundamental.

1. Plato, *Republic*.

While this orientation toward a deeper and more stable reality is compelling, it does not remain uncontested. Later philosophical developments introduce a different perspective on the relation between appearance and reality. Immanuel Kant [2], for example, argues that what we perceive is always shaped by the structure of the human mind. Reality, as it appears to us, is already organized through categories such as space and time, and what lies beyond this structure—the thing-in-itself—remains inaccessible.

This creates a fundamental tension. If reality is always mediated through perception, then the possibility of fully accessing it becomes uncertain. The distinction between what is and how it appears cannot be entirely overcome.

Georg Wilhelm Friedrich Hegel [3] responds to this tension by rejecting the idea of an inaccessible reality behind appearances. For him, reality is not divided in this way. Instead, it unfolds as a unified process in which thought and being develop together. What appears as separation is, in this view, part of a larger movement toward coherence.

Yet this movement toward system and unity is met with further resistance. Søren Kierkegaard [4] shifts attention away from abstract structures and back toward the individual, emphasizing that existence is not lived at the level of systems, but in concrete decisions. Friedrich Nietzsche [5], in turn, questions whether there is any fixed reality behind interpretation at all, suggesting that what we take to be truth may be inseparable from perspective and the forces that shape it.

What emerges from these perspectives is not a resolution of the question of reality, but an expansion of it. Different answers do not eliminate uncertainty; they reveal its depth.

This orientation is powerful. It suggests that beneath the shifting surface of things, there is a deeper structure that can be understood, and that human reason has the capacity—at least in

2. Kant, *Critique of Pure Reason.*
3. Hegel, *Phenomenology of Spirit.*
4. Kierkegaard, *Fear and Trembling.*
5. Nietzsche, *Gay Science.*

part—to access it. Knowledge, in this sense, is not simply the accumulation of observations, but a movement toward what does not change.

Yet this approach introduces its own difficulty. Even if such a deeper structure exists, it is not immediately clear how it can be accessed or verified. The claim to have reached a more fundamental level of truth raises further questions about interpretation, perspective, and the limits of human understanding. How do we distinguish between genuine insight and projection? How do we ensure that what we take to be universal is not shaped by our own assumptions?

The attempt to resolve uncertainty by appealing to a higher order does not eliminate the problem; it transforms it.

Science, by contrast, approaches the question from within experience rather than beyond it. It focuses on observation, measurement, and the development of models that describe how processes unfold. In doing so, it has achieved a remarkable degree of explanatory power. Cosmology, for example, provides detailed accounts of the evolution of the universe, from an early, dense state to the formation of galaxies, stars, and planetary systems.

These achievements are significant—not only because they expand our knowledge, but because they demonstrate that certain aspects of reality can be understood with precision and reliability. Science does not merely speculate; it produces explanations that can be tested, refined, and applied. It builds models that work, and those models allow us to interact with the world in increasingly sophisticated ways.

And yet, even here, the question persists.

Scientific explanations describe how the universe develops, but they do not fully resolve why it exists at all. The origin of the laws that govern it, the conditions that make its existence possible, and the question of why there is something rather than nothing remain open. At the edge of scientific inquiry, explanation gives way once again to uncertainty—not because science has failed, but because its domain is specific.

Science explains processes within reality; and yet it does not necessarily explain the existence of reality itself.

We are left in a position that is neither one of ignorance nor of complete understanding. Instead, we find ourselves within a tension: we possess partial insights, structured explanations, and powerful conceptual tools, yet we do not have a complete or final account.

This tension is not accidental. It reflects the condition in which human understanding operates. We are capable of knowing, but not of knowing completely. We can construct explanations, but not ones that fully close the question of existence.

Faced with this tension, there is a natural inclination to seek closure—to adopt systems that offer comprehensive explanations, or to commit to frameworks that provide stability and direction. Religion, philosophy, and even certain interpretations of science can fulfill this role. Each can present itself not merely as a perspective, but as a fully comprehensive account of reality.

Such systems are not without value. They provide orientation, enable action, and allow us to make sense of complex realities. Without them, the world would be difficult to navigate. But they also carry a risk. When a system is treated not as a way of engaging with reality, but as a final representation of it, it begins to close the very question it was meant to address.

This brings us back to the Human Condition.

Why is there something rather than nothing?

This question does not point toward a single answer, but toward different ways of engaging with reality. Each approach offers a partial understanding, shaped by its own perspective and limits. What remains is not a final conclusion, but an open landscape in which understanding continues to develop.

What began as an attempt to understand takes shape as a claim to finality.

The alternative is not to reject systems altogether, but to recognize their limits. Understanding is possible, but always partial. Explanation is valuable, but never complete. Meaning can be approached but not fully captured within a single framework.

This does not leave us without direction. It changes the nature of inquiry.

The question is no longer how to arrive at a final answer, but how to engage with a reality that exceeds any single system of explanation. It requires a shift from seeking closure to sustaining awareness, from final answers to ongoing engagement.

We do not stand outside the question of existence, observing it from a distance.

We are already within it.

And it is from within this condition—not beyond it—that any attempt at understanding must begin.

The question of existence does not end in abstraction. It takes shape in how we live, decide, and act.

Chapter 2

Philosophy, Good, and Evil

If the starting point is the recognition that we exist without fully understanding why, the next question follows almost inevitably: how, then, should we live? Unlike the question of existence itself, this appears at first to be more practical. It concerns action rather than origin, decision rather than explanation. And yet, it is no less complex, because embedded within it is a problem that has shaped philosophical reflection for centuries: the distinction between good and evil.

At an intuitive level, this distinction seems clear. We associate what is good with compassion, justice, and truth, and what is evil with harm, injustice, and deception. These associations provide a basic orientation that allows us to navigate the world, to make judgments, and to act with a sense of direction. Without some form of moral differentiation, action itself would become difficult, if not impossible.

Yet, this apparent clarity begins to dissolve when we move from abstract definitions to lived experience.

In concrete situations, moral decisions rarely present themselves in simple terms. What appears clear in principle often becomes less certain in practice—when decisions involve conflicting responsibilities, limited information, or consequences that cannot be fully anticipated.

Actions that appear justified from one perspective may reveal harmful consequences from another. Intentions are often mixed, shaped by motives that are not fully transparent even to the individual who acts. Outcomes cannot always be predicted, and what is meant as good may result in unintended harm. The distinction between good and evil, which seems stable in theory, comes to be more ambiguous in practice.

More importantly, this distinction does not lie entirely outside us. It is not simply a matter of evaluating the actions of others. The same individual is capable of both, often within the same context. The line between good and evil does not run between people, but through each of them.

This complicates the picture in a fundamental way. Good and evil are not merely external categories that can be applied to the world at a distance. They are possibilities that emerge within human action itself. We are not neutral observers who classify moral phenomena; we are participants whose decisions give those distinctions their concrete form.

Faced with this complexity, philosophical thought has often sought a more stable foundation. If moral judgments vary and conflict, perhaps there exists a deeper structure that can ground them. This is the direction taken by thinkers such as Plato [1], who argued that the instability of human perception points to a more enduring level of reality. In his view, true knowledge—including moral knowledge—must be rooted in something that does not change. The good, in this sense, is not merely what appears beneficial in a particular situation, but something that exists independently of shifting human perspectives.

The appeal of this position is clear. It suggests that morality is not arbitrary but grounded in something more stable than shifting human perspectives. It offers the possibility that moral truth can be discovered rather than constructed, and that human judgment can align itself with something more stable than individual opinion.

Yet this approach raises further questions. Even if such a foundation exists, how do we gain access to it? The claim that there

1. Plato, *Republic.*

is a higher moral order does not, by itself, resolve the problem of interpretation. It remains unclear how we distinguish between genuine insight into this order and projections shaped by culture, belief, or personal inclination. The attempt to secure morality through an absolute foundation risks introducing a different form of uncertainty—one that lies not in the absence of truth, but in the difficulty of recognizing it.

A different approach can be found in Aristotle, who does not ground morality in an abstract and unchanging ideal, but in practice. For him, ethical life develops through habit, experience, and the cultivation of character. What is good is not determined solely by fixed principles, but by the ability to act appropriately within specific situations. This shifts the focus from absolute foundations to practical judgment. [2]

While philosophy seeks structure, experience often reveals something less orderly.

The way in which truth and meaning are approached has shifted across different periods of thought. In ancient Greek philosophy, a transition took place from myth-based explanations toward inquiry grounded in observation and reasoning. Thinkers such as Aristotle explored the possibility that ethical understanding and fulfillment could be developed within human life itself, rather than derived from a transcendent source. This orientation continued in the Hellenistic schools, which emphasized practical ways of living, often without reliance on fixed dogma or a defined afterlife.

With the rise of Christianity, the focus moved more strongly toward a transcendent foundation of truth, providing stability and authority, but also reshaping earlier approaches. Figures such as Thomas Aquinas [3] later attempted to reconcile these perspectives by integrating Aristotelian reasoning into a theological framework. In modern philosophy, this tension re-emerges in thinkers such as Nietzsche and the existentialists [4], who again place responsibility

2. Aristotle, *Nicomachean Ethics.*
3. Aquinas, *Summa Theologica.*
4. Sartre, *Existentialism Is a Humanism.*

for meaning within the individual. Yet even this human-centered approach is not without limits, as contemporary insights suggest that the ways in which we construct truth are shaped by the conditions of human perception itself[5].

This becomes particularly visible in experience and is powerfully illustrated in the work of Fyodor Dostoevsky[6], where morality is not presented as a system of clear principles, but as a field of inner conflict. His characters do not simply act according to what they know to be right. They struggle, hesitate, justify, and contradict themselves. They can recognize moral truth and yet act against it.

Moral failure cannot be explained solely in terms of lack of knowledge. It points instead to a division within the individual. Human beings are not fully unified in their motivations. They are shaped by competing impulses, desires, fears, and rationalizations. As a result, the question of good and evil cannot be reduced to a problem of correct understanding. It is also a problem of how individuals relate to what they understand.

This inner division does not occur in isolation. It is shaped by social and biological conditions that influence how individuals perceive and act.

Human beings are deeply social. The need for belonging, recognition, and acceptance plays a central role in behavior. Individuals tend to align with group norms, adopt shared perspectives, and adjust their actions in response to social pressure. What is perceived as right or justified can therefore be influenced by the surrounding environment.

Historical examples make this visible. In situations where group identity comes to be dominant and alternative perspectives are suppressed, actions that would otherwise be questioned can come to be accepted or even encouraged. The events surrounding World War II illustrate how destructive outcomes can emerge not only from individual intent, but from collective dynamics in which conformity, fear, and identification override reflection.

5. Hoffman, *A case against reality.*
6. Dostoevsky, *Crime and Punishment.*

This does not imply that individuals are without responsibility. It shows that moral action is not formed in isolation, but within conditions that can amplify or diminish reflection.

Even so, biological and psychological processes influence behavior in ways that are not always transparent. Human action is connected to underlying mechanisms—such as reward systems, emotional responses, and learned patterns—that shape motivation. Desires, impulses, and forms of short-term satisfaction can influence decisions, sometimes in ways that conflict with longer-term considerations or stated values.

In modern contexts, these mechanisms are increasingly engaged by environments that stimulate desire and reinforce immediate gratification. Many forms of contemporary life are structured around fulfilling short-term needs and desires, making it more difficult to maintain distance from impulses that arise.

This dimension was already recognized in a different form by Søren Kierkegaard,[7] who described what he termed the aesthetic stage of life. In this mode, individuals are oriented toward experience, pleasure, and immediate satisfaction. While not inherently negative, this orientation can become limiting when it dominates, as it reduces the space for reflection and responsibility.

Contemporary psychology further complicates the picture. Research suggests that behavior is influenced by factors such as genetic predispositions, environmental conditions, and unconscious processes. In this view, actions are not always the result of fully autonomous choice but emerge from a complex interaction of influences.

As argued by Julia Shaw [8] in Making Evil, what is often labeled as "evil" may not always stem from deliberate intention, but from processes that are difficult to fully control. Impulses, situational pressures, and psychological mechanisms can lead to actions that individuals themselves might not fully understand or anticipate.

At the same moment, other perspectives maintain that responsibility remains meaningful. Philosophers such as Daniel

7. Kierkegaard, *Fear and Trembling*.

8. Shaw, *Making Evil*.

Dennett [9] argue that even within a structured and causal world, individuals can be understood as agents capable of reflection and self-regulation. Responsibility, in this sense, does not require complete independence from all influences, but the capacity to respond to them.

This creates a further tension. Human beings are influenced by forces they do not fully control, yet they are still held responsible for their actions. The distinction between good and evil is therefore shaped not only by abstract principles, but by the conditions under which individuals act.

This tension between knowledge and action can also be understood more broadly as a tension between different ways of relating to reality. Some approaches emphasize structure and coherence, seeking to understand moral life within a larger system. Others focus on the individual, where moral decisions are experienced not as elements within a system, but as immediate and often conflicting possibilities.

In more systematic perspectives, morality can appear as something that can be clarified through reason, integrated into a broader framework, and understood as part of a coherent whole. Yet from the standpoint of lived experience, this coherence is not always present. The individual does not encounter morality as a completed system, but as a series of decisions made under conditions of uncertainty.

This difference helps explain why moral understanding can remain unstable, even when it is conceptually well developed. The structure may be clear, but the experience of acting within it is not.

The stability of moral systems is challenged further by the critique of their historical and psychological foundations. Thinkers such as Friedrich Nietzsche [10] argued that moral values are not timeless truths, but developments that emerge within specific cultural contexts. According to this view, what is taken to be universally valid is often the result of historical processes, shaped by power, tradition, and perspective. Moral systems may reflect

9. Dennett, *Freedom Evolves.*

10. Nietzsche, *Gay Science.*

not only insight, but also the conditions under which they were formed.

This does not necessarily render all values arbitrary. But it does call into question the assumption that they rest on an unquestionable foundation. What appears as absolute may, in fact, be contingent. What is presented as universal may reflect a particular point of view.

The implications of this critique are significant. If moral systems are historically conditioned, then the authority they claim becomes less stable. What was once regarded as self-evident must now be examined. Yet this examination does not eliminate the need for moral judgment. Even if values are not grounded in a fixed and universal order, the experience of responsibility remains. We continue to act, to decide, and to live with the consequences of what we do.

This creates a tension that is difficult to resolve. On the one hand, the collapse of a fixed foundation introduces a degree of freedom. Individuals are no longer bound to accept moral systems as given. On the other hand, this freedom is accompanied by uncertainty. Without a stable reference point, the basis for moral judgment comes to be less clear, while the necessity of making such judgments does not disappear.

It is within this tension that Søren Kierkegaard [11] places the individual. For him, the question of how to live cannot be resolved at the level of abstract systems alone. It must be confronted within existence itself. Truth, in this sense, is not merely something to be understood intellectually, but something that must be lived.

This introduces a different dimension. The individual cannot fully rely on external systems to determine what is right. At some point, a decision must be made, and that decision is made without complete certainty. Responsibility cannot be transferred entirely to doctrine, reason, or tradition. It must be assumed.

What emerges from these perspectives is not a unified theory of morality, but a recognition of its complexity. The desire for a stable foundation remains, but it is accompanied by an awareness

11. Kierkegaard, *Fear and Trembling.*

of its limits. The experience of inner conflict challenges the idea that moral clarity can be achieved simply through knowledge. The critique of historical development undermines the assumption that moral systems are absolute. And the emphasis on individual responsibility highlights the fact that moral life cannot be fully delegated to any external structure.

In the face of this complexity, there is a strong temptation to simplify. It is often appealing to define good and evil in clear and absolute terms, to establish firm boundaries that eliminate ambiguity. Such simplification provides orientation and reduces uncertainty. However, it also carries a risk. When moral distinctions are fixed into rigid frameworks, they can lose contact with the complexity of lived experience. What was intended as clarity can become reduction.

This is where moral thought begins to move toward absolutism.

Not because the intention to clarify is misguided, but because the process of clarification develops into closure. The system is no longer open to revision or reflection. Questioning is treated not as part of understanding, but as a threat to it. The complexity of moral life is replaced by a structure that appears complete but may no longer correspond to the reality it seeks to describe.

However, the rejection of such systems does not provide a simple alternative. If all fixed foundations are abandoned, the risk emerges that moral distinctions lose their force altogether. If everything comes to be relative, then the basis for judgment becomes unclear, and responsibility risks being weakened rather than strengthened.

We are left in a position that resists simple resolution.

There is no final definition of good and evil that eliminates ambiguity. Nevertheless, these distinctions are not arbitrary. They are embedded in human action and experience. We encounter them not as abstract categories, but as dimensions of our own lives.

To recognize this is not to abandon the question, but to situate it more accurately.

Our experience does not stand outside morality, applying definitions from a distance.

We are always already involved.

Our actions, decisions, and relationships are the context in which the distinction between good and evil takes shape. It is not something fully determined before we act, but something that emerges through action itself.

If philosophy reveals the complexity of moral life and the difficulty of grounding it in a single framework, religion can be seen as one of the most influential attempts to provide such a framework.

The question, however, remains whether this structure resolves the problem—or reconfigures it in a different form.

Chapter 3

Religions Examined

Continuity, Development, and Interpretation

IF PHILOSOPHY REVEALS THE complexity of moral life and the difficulty of grounding it in a single framework, religion can be understood as one of the most influential attempts to provide such a framework. It not only addresses how we should live—it also seeks to answer the deeper question introduced at the outset: why there is something rather than nothing, and what it means that we are part of it.

In doing so, religion offers something that philosophy alone often does not: a structured sense of meaning. It situates human existence within a larger narrative, one that connects origin, purpose, and destiny. For many, this provides orientation and stability, particularly in the face of uncertainty. Religious traditions frequently present themselves not as tentative interpretations, but as revealed truths—claims that are not merely proposed but affirmed with authority.

Yet when these traditions are considered alongside one another, a fundamental tension emerges. They do not present a single unified account of reality, but multiple, sometimes conflicting ones.

Judaism, Christianity, and Islam each understand themselves as grounded in divine revelation. Each contains internally coherent systems of belief and practice. Each has developed rich intellectual traditions that engage with questions of ethics, metaphysics, and human existence. And yet, they differ not only in interpretation, but in their most fundamental claims about revelation, authority, and the nature of truth.

This plurality introduces a further difficulty. If one of these traditions is true in an absolute sense, what does that imply about the others? Conversely, if more than one contains truth, how are their differences to be understood? Historically, this tension has often been resolved through exclusivity. A tradition affirms itself as the final or most complete expression of truth, while others are reinterpreted, subordinated, or rejected.

Such a resolution provides clarity, but it does so by simplifying a more complex reality.

To understand that complexity, it is necessary to look beyond the idea that these traditions emerged independently. When placed within a broader historical context, they reveal patterns of continuity. Earlier systems, such as Zoroastrianism [1], introduced key elements that later reappear in transformed forms: linear time, a moral tension between good and evil, the idea of judgment, and the expectation of a final resolution. These elements do not remain unchanged, but they shape the framework within which later monotheistic traditions develop.

Zoroastrianism offers an example of how influential frameworks of meaning can emerge, shape civilizations, and later recede from view. For centuries, it provided a structured approach to questions of good and evil, the balance between the material and the spiritual, and the role of human responsibility in shaping both. This orientation is reflected in early Zoroastrian texts, which speak of "the joy of both worlds, the physical and that of the mind," [2] suggesting an attempt to relate ethical life to both human experience and a broader, enduring order.

The interaction between Persian and Jewish traditions during periods of exile suggests that such ideas do not develop in isolation,

1. Boyce, *Zoroastrians.*

2. Kaizai, ,*Gathas,* 50.

but through contact and exchange. Yet despite its long-standing influence, Zoroastrianism gradually faded from prominence as other religious frameworks became dominant, only to be rediscovered much later. This trajectory illustrates how ways of understanding truth can persist, transform, or disappear within shifting historical and cultural conditions, shaped not only by reflection, but also by the broader forces within which they are sustained.

Earlier traditions also reveal that many of the structures later associated with monotheistic religion were not entirely new. In ancient Egyptian [3] belief, for example, one finds a complex integration of creation, death, and renewal within a unified framework. Different aspects of the divine—associated with creation, judgment, and continuation—are not separate in an absolute sense but form a structured whole.

This reflects an early attempt to understand existence through interconnected dimensions of origin, transformation, and continuation. These are attempts that predate later religious systems. Even so, these earlier traditions often emphasize cyclical patterns of life and renewal, in contrast to the more linear structure that takes shape as characteristic of later monotheistic thought.

This historical perspective also highlights a broader distinction between different cultural approaches to time and existence.

This distinction is not merely theoretical—it shapes how finality itself is understood. A linear conception of time lends itself to the idea of a definitive conclusion—an ultimate resolution in which truth is fully revealed. A cyclical conception, by contrast, resists such closure. It emphasizes continuity and transformation rather than final judgment.

In this sense, the structure of time itself influences the tendency toward absolutism. Where reality is understood as moving toward a single endpoint, the idea of a final and complete account emerges as more compelling. Where reality is understood as ongoing, such closure emerges as more difficult to sustain.

The monotheistic traditions that developed in the Middle East tend to adopt a linear view of time: a beginning, a progression, and

3. Assmann, *Search for God in Ancient Egypt*.

an end. History moves toward fulfillment, judgment, or resolution. This structure reinforces the idea that there is a definitive account of reality toward which everything is directed.

In contrast, earlier Greek thought did not emphasize a single beginning in the same way. Nor do traditions such as Hinduism and Buddhism, which often understand existence in cyclical terms. In these frameworks, reality is not moving toward a final endpoint, but unfolding through recurring processes of emergence, transformation, and return.

This difference is not merely conceptual – it shows that even the structure of time influences the movement toward stabilization and closure. It shapes how truth is understood. A linear framework lends itself more easily to the idea of a final revelation or ultimate conclusion. A cyclical framework, by contrast, often resists closure, emphasizing continuity rather than finality.

The structure of time itself thus influences how closure and finality are approached.

The development of religions is not a single, completed process. It unfolds over time, shaped by transmission, interpretation, and interaction. Rather than emerging all at once, traditions take form across different periods, influenced by many voices, texts, and contexts.

Even Judaism, for example, developed significantly between 516 BCE to 70 CE, incorporating apocalyptic ideas, evolving messianic expectations, and elements of mystical thought. As scholars such as John J. Collins have noted, apocalyptic traditions are not only concerned with end times, but also with processes of transformation and renewal [4]. Christianity in the first century is not the same as in the fourth and fifth century, reflecting similar processes of development and reinterpretation.

This continuity suggests that religions are not isolated revelations appearing in a vacuum, but part of an evolving attempt to understand existence. They inherit structures, reinterpret them, and adapt them to new contexts. What appears as absolute and

4. Collins, *Apocalyptic Imagination.*

final within a tradition often has a longer and more complex development behind it.

Meanwhile, religions do not simply transmit ideas—they organize them.

Texts are collected, interpreted, and eventually fixed into canonical forms. This process is particularly visible in early Christianity, where a variety of writings circulated among different communities. These writings did not all present the same understanding of the message attributed to Jesus. Some emphasized faith and belief, while others placed greater emphasis on insight and inner transformation.

Within early Christianity itself, differences in emphasis can be observed that further illustrate the development of interpretation. Some strands place strong emphasis on ethical practice and the way one lives in relation to others, while others focus more explicitly on belief, salvation, and the role of faith.

These differences are not merely variations in expression but reflect distinct orientations. In some cases, the message centers on transformation through action and responsibility; in others, it is structured around belief in a defining event and its significance. Over time, one of these orientations comes to be more dominant, as it allows for greater doctrinal clarity and institutional stability.

In practice, these differences are not only theoretical. They shape how people live, how communities organize themselves, and how meaning is experienced in everyday life.

This development does not necessarily resolve the underlying diversity. Rather, it organizes it. What was once a field of multiple interpretations takes shape as a more defined system, in which certain perspectives are emphasized while others recede.

Over time, this diversity was reduced through processes of selection. Certain texts were recognized as authoritative, while others were excluded. Among those excluded is the Gospel of Thomas [5], a collection of sayings that presents a markedly different emphasis from the canonical tradition.

5. Meyer, *Nag Hammadi Scriptures.*

Rather than focusing primarily on belief, it directs attention inward. This difference becomes particularly visible in the following statement (Gospel of Thomas, Logion 70):

"If you bring forth what is within you, what you bring forth will save you. If you do not bring forth what is within you, what you do not bring forth will destroy you." [6]

Here, the emphasis shifts more often toward the individual. Truth is not simply something to be accepted; it is something to be realized. The responsibility lies not only in believing, but in bringing forth what is already present within oneself. This introduces a dimension of inward development that is less dependent on external authority.

In contrast, canonical texts such as the Gospel of John emphasize a different orientation. In John 20:29, it is stated:

"Blessed are those who have not seen and yet have believed."[7]

Here, faith emerges as central. The relation to truth is not grounded in direct realization, but in trust—trust in what cannot be fully verified through experience. This makes possible a shared religious framework in which belief can function as a unifying principle across a community.

These two orientations are not necessarily incompatible, but they highlight a tension within religious development. One emphasizes inner realization; the other emphasizes shared belief. As religious traditions become institutionalized, the latter often emerges as more dominant, as it allows for structure, continuity, and collective identity.

The process through which religious texts became authoritative is not only theological, but also historical and contextual. Scholars such as Bart Ehrman [8] have shown that many biblical texts reflect layers of editing, variation, and interpretation over time. Rather than emerging as a single, unified account, they were

6. Pagels, *Beyond Belief*, 53. Translation by MacRae, less lucid, offered in class, 202.

7. John 20:29

8. Ehrman, *Misquoting Jesus*.

shaped by different communities, each with its own concerns, perspectives, and theological emphasis.

This does not necessarily diminish their significance. But it does challenge the idea that they represent a fixed and internally consistent whole. What appears as a unified tradition is, in many cases, the result of selection, transmission, and interpretation.

Even within the New Testament, there is no complete agreement on fundamental aspects. The Gospel of John, for example, presents a more developed theological perspective than the earlier synoptic gospels. It emphasizes the divine nature of Christ in a way that appears more explicit and structured, suggesting that theological reflection evolved over time.

And yet, this development is itself subject to debate. Scholars such as George van Kooten have proposed alternative interpretations of the Gospel of John, illustrating that even its historical position continues to be subject to debate. Van Kooten [9] argued that John may in fact represent an earlier strand of thought from earlier than 70 CE, challenging the assumption that it is simply a later theological construction. Others like Bart Ehrman [10] and many other scholars disagree on this view and argued it is between 90 to 110 CE. This lack of consensus illustrates a broader point: even the historical structure of foundational texts remains open to interpretation.

What emerges is not a fixed system, but a field of ongoing interpretation.

A similar complexity appears in Islam. The Qur'an presents itself not as an entirely separate tradition, but as a continuation and clarification of earlier revelations. It affirms many of the same figures and themes found in Judaism and Christianity, while also offering its own framework.

At the same time, it emphasizes a theme that resonates across these traditions: the unity of humanity and the importance of moral responsibility. In Surah 49:13 [11], it is stated that humanity

9. Kooten, *Reverberations of Good News.*

10. Ehrman, *Gospel of John.*

11. Qur'an 49:13

has been created in diversity "so that you may know one another," and that the most honored are those who are most conscious of their moral responsibility.

This reflects a dimension that is often overlooked when religions are compared superficially. Despite their doctrinal differences, they share underlying concerns: how to live, how to relate to others, and how to orient oneself toward something greater than individual existence.

However, in contemporary discourse, these similarities are frequently overshadowed by simplifications. Religions are reduced to their most visible or extreme expressions. Islam may be associated primarily with conflict, Christianity with institutional power, and other traditions with isolated characteristics that fail to capture their internal diversity.

These simplifications are not entirely without basis, but they remain incomplete. When they become fixed, they begin to function as a form of interpretation that resembles the very absolutism they may seek to critique. A complex system is reduced to a single defining feature, and that feature is treated as representative of the whole.

Meanwhile, it must be recognized that religions do not exist independently of human contexts. They develop within societies, interact with political structures, and are shaped by historical conditions. Institutions form, authority is established, and interpretations become standardized. The effect is that this gives a tension between the original impulse of a tradition—often rooted in insight or experience—and its later institutional form.

Across these developments, a recurring pattern becomes visible: a tradition begins with insight, develops structure, and gradually moves toward a more complete and authoritative form. This movement is not inherently problematic, but it carries a risk. The more a system presents itself as complete, the more difficult it comes to be to question its assumptions.

This is where the tendency toward absolutism reappears—not as a feature of one particular religion, but as a structural possibility within all of them. The desire for clarity, coherence, and certainty can lead to the closure of inquiry. What began as an attempt to understand develops into a system that defines the limits of what can be understood.

Recognizing this does not require the rejection of religion. Religious traditions continue to offer meaning, ethical guidance, and frameworks for understanding existence. But they must be approached with an awareness of their development, their diversity, and their limits. No single tradition fully captures what it seeks to describe.

When placed side by side, what becomes visible is not simply contradiction, but a shared effort. Each represents a different way of engaging with the same fundamental questions. Their differences highlight their limits; their similarities point to something that exceeds them.

Historical Development and Transformation of Religious Structures

To understand this continuity more fully, it is necessary to look more closely at how these structures emerge across different traditions.

In Ancient Egypt (around 2500 BCE), we already encounter a structured view of existence in which creation, judgment, and continuation are interconnected.[12] While this early framework focused on moral justification before the gods to enter the fields of Reeds, it laid a structural foundation for what would later become the iconic "weighing of the heart" against the feather of Ma'at—representing truth and order—determines whether an individual enters the Field of Reeds or ceases to exist. Here, moral evaluation is present, but it coexists with ritual elements and symbolic structures that allow for a more fluid interaction between order, fate, and intervention.[13]

12. Allen, *Ancient Egyptian Pyramid Texts*, 123-25.

13. Hornung, *Books of the Afterlife*, 13–15.

This indicates that the attempt to relate human life to a larger structure of meaning predates later monotheistic systems.

In Zoroastrianism [14], this structure emerges as more explicitly moralized. Human beings are judged based on thoughts, words, and deeds, and the figure of Daena reflects one's own conscience. The crossing of the bridge is not dependent on ritual or status, but on ethical orientation. The introduction of a more defined distinction between good and evil, combined with the expectation of a final restoration (Frashokereti), marks a significant development toward structured moral accountability.

While Zoroastrianism introduces a more explicit moral structure, it does not necessarily lead to the idea of eternal punishment. The consequences of one's actions are real and significant, but they are not understood as final in an absolute sense. The concept of Frashokereti points toward an eventual restoration in which the world is renewed and purified.

In this framework, evil is not sustained indefinitely but ultimately overcome. Judgment remains meaningful, but it is situated within a broader process rather than a final and irreversible division.

At this point, the question arises whether later traditions emerge independently, or whether earlier frameworks are reinterpreted within new historical and cultural contexts.

This question becomes particularly visible in shared narratives such as the flood story. In Mesopotamian traditions, the flood is often associated with disturbance or imbalance within a world governed by multiple divine forces. Survival is tied to preparation, preservation, and practical continuity. In the version found in Genesis, the structure shifts toward moral interpretation. The flood takes shape as a response to corruption and violence, and the narrative is embedded within a covenantal framework that emphasizes responsibility and renewal.

This leads to a further question of why these narratives take different forms. The differences are not only structural, but contextual. The version found in Genesis emerges within a specific

14. Boyce, *Zoroastrians.*

historical setting, often associated with the period of the Babylonian exile, in which questions of identity, continuity, and meaning become particularly urgent.

In such a context, narrative does more than describe events. It contributes to the preservation and redefinition of a collective identity. The emphasis on moral order, covenant, and responsibility can be understood not only as theological development, but also as a way of maintaining coherence under changing and uncertain conditions.

A similar dynamic can be observed in the recurring role of Egypt within biblical narratives. Egypt appears not only as a geographical location, but as a symbolic reference point—associated with origin, displacement, and transformation. Its presence across different texts suggests that religious narratives may incorporate earlier cultural elements, reinterpreting them within new frameworks of meaning.

This does not imply a single explanatory cause. Rather, it indicates that religious development is shaped by multiple factors, including historical experience, cultural interaction, and the need to establish continuity over time.

The transformation is not merely narrative, but structural. What was once part of a broader cosmological setting comes to be aligned with moral order and purpose.

It doesn't provide a single explanation for how these traditions are related. But it does suggest that what appears as distinct and self-contained may, in fact, reflect processes of reinterpretation, adaptation, and development over time.

What emerges from these patterns is not a reduction of religious traditions to a single origin, nor a dismissal of their distinctiveness.

Rather, it suggests that they operate within a shared field—one shaped by recurring human questions about existence, order, and meaning, in relation to a reality that is not fully accessible.

Within this field, different traditions develop their own narratives, symbols, and structures. These may appear unique when viewed in isolation yet reveal points of convergence when considered across time and context.

Recognizing these patterns does not eliminate difference, but it introduces nuance into claims of absolute uniqueness or superiority.

It becomes possible to see religions not only as competing accounts of truth, but as distinct responses to questions that remain, in many respects, open.

If religious traditions attempt to provide a structured account of meaning and reality, the question persists whether such structure can be grounded beyond interpretation alone.

Chapter 4

Toward a Position

Between Explanation, Meaning, and Absolutism

Religious Interpretations on Good and Evil

Where do we stand when we consider good and evil, and the persistence of suffering in the world—both ever-present?

The major religious traditions have not avoided this question. They have attempted to interpret it.

Within the three Abrahamic religions, different frameworks emerge.

In Judaism, as reflected for example in the Book of Job [1], evil is not presented as an independent force equal to God. Rather, it appears as something that operates within a structure permitted by God. Satan functions as an agent within a larger structure, and suffering points to a form of testing. Human responsibility remains intact, even within this framework.

In Christianity, evil is often understood as entering the world through the misuse of human freedom. The narrative of Adam and Eve introduces the knowledge of good and evil as a turning point. Suffering and sin are then interpreted within a broader framework of redemption, in which the life and crucifixion of Jesus provide a

1. Job 1:12, 2:6.

path toward restoration. Within this perspective, suffering can also be seen as contributing to moral development or spiritual growth.

In Islam, evil is primarily associated with human action and responsibility. While God permits events to occur, evil is not attributed to divine intention. Suffering is often interpreted as a trial—an opportunity for growth, purification, and moral evaluation. As expressed in the Qur'an (67:2): "He who created death and life to test you as to which of you is best in deed." [2] Life and death are thus understood as a means of moral evaluation. At the same moment, the full reasoning behind events is considered beyond complete human understanding.

Across these traditions, a common challenge appears: how to reconcile the existence of evil and suffering with the idea of a meaningful or ordered reality.

However, these traditions must also confront a more difficult question: why does evil continue to occur within the very systems that seek to explain or overcome it? Historical and institutional realities show that harm, injustice, and suffering are not external to religion alone, but can also emerge within it.

This raises further questions.

Is evil best understood as a test, a consequence, or a mystery within a larger order?

Or do such explanations serve, at times, to preserve a particular framework?

Is suffering meaningful, or does meaning arise from how it is interpreted?

These questions do not yield a single answer.
They point instead to the limits of interpretation.

2. Qur'an, 67:2.

Scientific Perspectives on Behavior and Causation

From a scientific perspective, the question of evil is approached differently.

Rather than beginning with meaning, science examines patterns of behavior, biological mechanisms, and social dynamics.

From an evolutionary standpoint, moral tendencies such as empathy, fairness, and cooperation can be seen as developments that support survival within groups. These tendencies help regulate behavior, reduce conflict, and increase collective stability.

At the same time, the same evolutionary logic allows for behaviors that we would describe as harmful or destructive. Aggression, competition, and exploitation can provide short-term advantages under certain conditions. In this sense, what is perceived as "good" and "evil" may emerge from the same underlying processes.

Historical analysis adds another layer. Group identity can lead to division between "us" and "them," reducing moral concern for those outside the group. Under such conditions, actions that would otherwise be rejected—such as violence, exclusion, or domination—can become justified within a collective framework.

Neuroscience introduces further complexity. Brain processes involved in judgment, empathy, and impulse control can vary between individuals. Factors such as trauma, environment, or neurological differences can influence behavior in ways that are not fully under conscious control. Certain conditions—such as psychopathy or severe impulse disorders—are associated with altered patterns in brain function that affect moral behavior. [3]

This does not eliminate responsibility; it complicates it. As noted earlier in the work of Julia Shaw [4], behaviors often associated with 'evil' can emerge from complex psychological and situational processes rather than from clear, deliberate intent.

Across both religious and scientific perspectives, a similar tension remains: human beings appear to act, to choose, and to be

3. Damasio, *Descartes Error.*

4. Shaw, *Making Evil.*

responsible, yet their actions are influenced by factors that are not fully transparent or controllable.

No single framework fully resolves this tension.

At this point, a transition becomes necessary.

If both religion and science attempt to address the same underlying reality—each from a different perspective—then the question is no longer which one replaces the other.

It develops into:

How should these perspectives be understood in relation to one another?

If religious traditions offer interpretations of good and evil, and philosophical reflection exposes the limits of certainty within those interpretations, science introduces a different way of approaching the same underlying questions.

It does not begin with narrative or belief, but with observation, measurement, and method. Its strength lies in its ability to produce explanations that are testable, reproducible, and increasingly precise.

This success is undeniable. Modern science allows us to describe the evolution of the universe, to understand the mechanisms of life, and to intervene in the material world with remarkable effectiveness. It does not merely interpret reality—it enables us to act within it.

This practical effectiveness creates a strong sense of reliability. Over time, this reliability can extend beyond method into interpretation, where what proves effective in explaining parts of reality begins to be understood as indicative of reality as a whole.

Yet it is precisely this success that introduces a new tension.

Science begins as a method, but it is often carried into a broader worldview. What starts as an approach to understanding specific aspects of reality gradually can come to function as a broader claim about reality as a whole. The distinction between explanation and interpretation becomes less clear. What can be measured comes to be what is considered real; what cannot be measured risks being dismissed as secondary or illusory.

This blurring does not occur as a deliberate shift, but as a gradual extension. Models that successfully describe particular processes gain authority, and that authority can be carried into areas where the model itself was not originally designed to apply. In this way, the boundary between what is explained and what is assumed takes shape as less clearly defined.

This movement is subtle, but significant. It mirrors, in a different form, the tendency already observed in religion: the movement from partial insight toward claims of completeness.

This tendency becomes visible in certain strands of scientific thought that adopt a strongly reductionist position. In such views, complex phenomena are explained entirely in terms of their simplest components. Life comes to be the expression of genetic mechanisms; consciousness takes shape as a byproduct of neural processes; meaning develops as something derivative, without independent standing.

Thinkers such as Richard Dawkins [5] have been influential in articulating this perspective, particularly in relation to evolutionary biology. His work has played an important role in challenging religious dogma and defending the importance of rational inquiry. His critique is often precise and, in many contexts, necessary, especially where belief systems resist evidence.

This reflects a broader pattern. A model that is highly effective within a specific domain can come to shape expectations about other domains. The success of explanation can encourage the assumption of completeness. What begins as a powerful tool develops into, almost imperceptibly, a framework through which reality itself is interpreted.

However, his position does not remain purely critical. It evolves into a broader interpretation in which the explanatory success of evolutionary theory is extended into a more comprehensive account of life itself. In this view, genes become the primary drivers of biological reality, and higher-level phenomena are understood largely as consequences of underlying mechanisms.

5. Dawkins, *Selfish Gene*.

What is important here is not that this view is without merit, but that it tends toward closure. It presents itself not merely as one model among others, but as the most complete and authoritative account available.

Yet this is precisely where tension re-emerges—this time from within science itself.

Scientific perspectives on Biology

Biologists such as Denis Noble [6] challenge the idea that life can be fully understood through a gene-centered framework. While not rejecting evolutionary theory, Noble argues that the reduction of biological systems to genetic determinism overlooks the complexity of how organisms actually function. Genes do not operate in isolation; they are embedded in networks of interaction in which higher-level structures influence lower-level processes.

What comes to be visible here is not a failure of science, but its internal diversity. Different models do not necessarily cancel each other out; they reveal that explanation operates at multiple levels. The attempt to reduce all phenomena to a single explanatory layer overlooks this plurality and risks simplifying what is inherently complex.

Causation, in this view, is not strictly bottom-up. It involves feedback across multiple levels, from cells to tissues to the organism as a whole. Biological systems are dynamic and context-dependent, and their behavior cannot be fully captured by focusing on a single explanatory level.

This does not invalidate science. It deepens it.

But it also reveals something important: within science, there is no single, uncontested framework. Even in fields that appear well established, competing interpretations exist. The idea that one model fully captures reality emerges as difficult to sustain when alternative models emerge from within the same empirical domain.

6. Noble, *Music of Life.*

Scientific perspectives on Cosmology and the Limits of Explanation

A similar tension appears in cosmology.

The Big Bang model provides a powerful account of the development of the universe, supported by extensive observational evidence. It explains the expansion of space, the formation of cosmic structures, and the distribution of matter and energy. And yet, it does not fully resolve the question of origin. What preceded the initial state, or what conditions made it possible, remains uncertain.

In this context, different interpretations arise. The boundary between explanation and origin marks an important limit. Scientific models can describe how the universe evolves from a given state, but the conditions under which such a state becomes possible remain less clearly defined. The question does not disappear; it shifts to a different level.

Some thinkers, such as Stephen Meyer [7], argue that the apparent fine-tuning of the universe and the complexity of life point beyond purely naturalistic explanations. They interpret scientific findings as suggestive of an underlying intelligence or design. From this perspective, the limits of current explanations are not merely gaps to be filled, but indications that the explanatory framework itself may be incomplete.

The shift from scientific observation to metaphysical conclusion is not unique to this perspective. It reflects a broader tendency to extend explanation into interpretation. The same data that supports a model of development can also be used to support a more comprehensive claim about the nature of reality.

Others, including researchers like Niayesh Afshordi together with a strong communicator like Phil Halper, take a different approach [8]. Rather than extending uncertainty into metaphysical conclusions, they emphasize the provisional nature of cosmological models. They explore alternative scenarios, question assumptions,

7. Meyer, *Return of the God Hypothesis.*
8. Halper and Afshordi, *Battle of the Big Bang.*

and highlight the limits of current knowledge without treating those limits as evidence for a specific external explanation.

This difference in approach highlights an important distinction. Uncertainty can be treated either as a problem to be resolved through a more comprehensive explanation, or as a condition that remains open within the process of inquiry. The direction taken influences not only conclusions, but the structure of thought itself.

Scientific and Religious perspectives

A similar dynamic appears in the dialogue between scientific and religious perspectives. Thinkers such as John Lennox [9] have sought to integrate scientific insights into a theistic framework, often in response to critiques of religion. At times, this can lead to interpreting scientific developments as confirmation of an already established metaphysical framework. While such approaches aim at coherence, they also raise the question of whether interpretation remains open—or moves toward reinforcement of a prior framework.

The attempt to integrate domains often arises from a desire for coherence. It seeks to avoid fragmentation by connecting different forms of understanding within a unified framework. Yet this same movement can lead to a form of closure, where interpretation comes to be aligned with a particular conclusion rather than remaining open to revision.

While such approaches aim to bridge domains, they can also reflect a tendency toward definitive interpretation. Scientific findings are not only engaged with but positioned as supporting a particular metaphysical conclusion. In this way, the effort to reconcile science and religion can mirror the same pattern observed elsewhere: partial insight extended toward a more comprehensive claim.

9. Lennox, *God's Undertaker*.

This does not invalidate the argument, but it raises the question of whether the interpretation remains open—or moves toward closure.

The contrast here is not simply between belief and disbelief. It is between different ways of interpreting uncertainty.

One approach treats the limits of explanation as signs pointing beyond the natural world. The other treats them as areas for continued investigation within it. Both operate within the same body of evidence yet arrive at different conclusions. This demonstrates that scientific data alone does not determine interpretation. It must be situated within a broader conceptual framework.

Data provides structure, but it does not dictate meaning. The interpretation of data depends on the conceptual frameworks within which it is situated. These frameworks guide what is considered relevant, how connections are made, and which conclusions are drawn.

At this point, an important insight emerges: science, like religion and philosophy, does not exist without interpretation. It is not purely neutral. It operates within frameworks that shape how evidence is understood and what conclusions are drawn from it.

This in no way undermines science—it clarifies its nature.

The problem arises when interpretation is no longer recognized as such.

When a particular scientific framework is treated not as a model, but as a complete description of reality, it begins to take on a role similar to that of doctrine. The distinction between method and worldview collapses, and what began as a powerful tool for understanding develops into a comprehensive claim about what exists and what does not.

This is where the critique of modern absolutism comes to be most visible.

Just as religious systems can become rigid when they present themselves as final and unquestionable, scientific perspectives can become rigid when they extend beyond their methodological limits. The assertion that reality is fully captured by a particular model—whether biological, physical, or cosmological—introduces a form of closure that may not be justified by the evidence itself.

This tendency is not confined to scientific theory. It also appears in contemporary intellectual discourse.

Thinkers such as Sam Harris [10] have argued strongly against religious belief, particularly in its more dogmatic forms, and have emphasized the role of reason and science in understanding the world. While such critiques can be valuable, they can also reflect a particular cultural perspective. When complex traditions are reduced to their most problematic aspects, and alternative viewpoints are dismissed rather than examined, the critique risks becoming reductive.

Scientific perspectives on History

A similar dynamic can be observed in historical interpretations of Western development. Writers such as Tom Holland [11] have highlighted the profound influence of Christian thought—particularly Pauline ideas—on the formation of Western values. While this perspective brings important continuities into focus, it can foreground certain influences while leaving others less visible, including the enduring impact of Greek philosophical traditions and the transmission and development of knowledge through Islamic scholarship. This raises a broader question of how historical narratives are shaped: whether they remain open to multiple strands of influence, or tend toward a more unified explanatory account.

The question, then, is not which perspective is final, but how these perspectives relate—and how we position ourselves within their tension.

10. Harris, *End of Faith*.

11. Holland, *Dominion*.

Patterns Across Perspectives

In each of these cases, the issue is not the presence of insight, but the movement toward exclusivity. A perspective that begins as a contribution to understanding can become a framework that claims primacy.

The pattern that emerges across religion, science, and philosophy is therefore not one of isolated error, but of a recurring structural tendency. Partial understandings, when taken beyond their appropriate scope, become candidates for absolutism.

Recognizing this does not require rejecting science, any more than it requires rejecting religion or philosophy. Scientific knowledge remains indispensable for understanding the structure of the world we inhabit. Its methods provide a level of clarity and reliability that cannot be ignored.

But it must be held within its limits.

At this point, the question returns—not to abstract systems, but to the individual engaging with them.

If no single framework provides a complete account, how should one relate to them? The answer cannot lie in simple rejection, nor in uncritical acceptance. It requires a form of engagement that recognizes both the strengths and the limits of each perspective.

This brings us back to a theme already encountered in earlier chapters. The emphasis on inward awareness, as expressed in the Gospel of Thomas, and the existential focus of Søren Kierkegaard [12], points to a dimension that cannot be fully captured by external systems. However, the insights of science remain essential for understanding the world in which that inward awareness takes place.

Neither can replace the other.

Positions that present themselves with clarity and certainty are easier to communicate and more readily shared. They require less explanation, leave less room for ambiguity, and are more easily shared within groups. In this sense, absolutism is not only a philosophical position—it is also a communicative advantage.

12. Kierkegaard, *Concept of Anxiety*.

Yet what is easy to communicate is not always what is most accurate. The reduction of complexity into clear and stable forms makes ideas more effective, but also more limited. What is gained in clarity may be lost in depth.

The task, then, is not to resolve the tension between these domains, but to remain within it without prematurely closing it. To recognize that explanation, meaning, and interpretation each play a role, but none is sufficient on its own.

If neither philosophical systems, religious frameworks, nor scientific explanations fully resolve the question of reality, then the issue is not which one is correct in isolation, but how these different ways of understanding relate to one another—and how we position ourselves within them.

The question develops into how one can live within their tension without turning them into absolutes.

Chapter 5

Living Without Absolutes

IF THE PREVIOUS CHAPTERS have shown anything, it is that we do not lack systems of understanding. On the contrary, we are surrounded by them. Philosophy offers frameworks through which we attempt to think clearly about reality and our place within it. Religion provides narratives that situate human existence within a broader structure of meaning. Science develops models that describe and predict the processes governing the physical world.

This position does not arise from indecision, but from recognition: an awareness that the alternatives—complete certainty on the one hand and complete relativism on the other—both fail to account for the conditions under which understanding actually develops.

Certainty promises stability, but often at the cost of excluding complexity. Relativism preserves openness, but risks dissolving distinctions that remain meaningful in practice. Neither position, taken in isolation, fully reflects the way human understanding functions.

Each of these domains has achieved something significant. Each reveals aspects of reality that would otherwise remain obscure. Each provides orientation in a world that would be difficult to navigate without them.

And yet, none of them is exhaustive.

This is not a conclusion reached lightly. It follows from the recognition that each system operates within a particular scope. Philosophy clarifies concepts but does not resolve existence. Religion offers meaning but develops historically and institutionally. Science explains mechanisms but does not fully account for why there is something rather than nothing. Their strengths are real, but so are their limits.

The difficulty lies in how we respond to this situation.

There is a persistent desire to arrive at a final ground—a point at which uncertainty ends and understanding comes to be stable. This desire is not accidental. It reflects a need for orientation, for clarity, and for the ability to act without constant hesitation. In the absence of such a ground, the world can appear unstable, and decision-making can become more demanding.

It is therefore understandable that individuals and societies gravitate toward systems that promise certainty. Religious traditions may present themselves as final revelations. Philosophical systems may claim to establish foundational truths. Scientific perspectives may extend their explanatory success into claims about the total nature of reality.

Each of these movements offers something valuable. But each also carries a risk.

The risk is not that these systems exist, but that they are taken to be complete.

When this happens, the system is no longer treated as a way of engaging with reality, but as a final account of it. What began as an attempt to understand develops into a structure that defines the limits of what can be understood. Questioning is no longer part of the process, but a threat to it.

This is the movement toward absolutism.

It does not belong to one domain alone. It appears in religion when doctrine emerges as unquestionable, in philosophy when systems claim completeness, and in science when models present themselves as complete representations of reality rather than as provisional explanations. The forms differ, but the structure is the same: partial understanding extended into total certainty.

Opposed to this is another position, often presented as an alternative: relativism. If no system can claim completeness, it may seem that all perspectives are equally valid, that there is no truth, and that distinctions between better and worse interpretations lose their meaning.

But this conclusion does not follow.

The recognition that no system is absolute does not imply that all perspectives are equivalent. Some explanations are more coherent, more consistent with evidence, and more responsive to reality than others. Some interpretations illuminate more than they obscure. The absence of final certainty does not eliminate the possibility of truth; it changes the conditions under which truth must be approached.

This creates a position that is more demanding than either absolutism or relativism.

It calls for thinking, judging, and acting without relying on a final guarantee. It requires that we remain open to revision without abandoning commitment. It requires that we engage with systems of thought while recognizing their limits.

Systems are necessary because they provide orientation. They allow individuals and communities to act, to decide, and to communicate within a shared structure. Without some degree of structure, the world would appear fragmented and difficult to navigate.

At the same time, these structures remain partial. They are constructed within specific contexts, shaped by particular perspectives, and limited by the conditions under which they emerge. Recognizing their necessity does not require treating them as complete.

In this sense, responsibility emerges as more direct.

We cannot fully defer to external authority. We cannot simply apply a fixed set of rules that resolves every situation. Nor can we retreat into the idea that no judgment is possible. We are required to decide, and those decisions have consequences—not only for ourselves, but for others.

This returns us to a fundamental point: we are not observers of reality, but participants in it. Our actions are not neutral. They shape the world in which we and others live. The question of how to live is therefore not an abstract problem that can be solved in advance. It unfolds within the very process of living.

At this point, the role of awareness comes to be central.

If no system can fully determine what is right, then the ability to reflect on one's own assumptions emerges as essential. This includes recognizing when a belief is being treated as absolute, when a perspective is being extended beyond its limits, and when complexity is being reduced in ways that obscure rather than clarify.

This form of awareness does not provide fixed answers. It does not eliminate uncertainty. But it changes how we relate to it. Instead of attempting to close the question, we remain within it more attentively.

Uncertainty, in this sense, is not simply a lack of knowledge. It is a structural feature of the relationship between understanding and reality. It persists not because we have failed to resolve it, but because it cannot be fully resolved without reducing what exceeds our frameworks.

What appears at this point is that understanding is not a fixed state, but an ongoing activity. It involves movement rather than arrival, adjustment rather than completion. Positions are formed, but they are not final. They remain open to revision as new perspectives and insights emerge. As understanding stabilizes, it comes to be easier to communicate and apply, but also more resistant to change.

To live without absolutes, then, does not mean to live without direction.

It means to act while acknowledging that understanding is incomplete.

It means to commit without claiming final certainty.

It means to hold beliefs without turning them into unquestionable truths.

This is not an easy position to maintain. It requires continuous engagement, and it resists the comfort of simple answers. But it may be the only position that remains faithful to the complexity of the situation we are in.

Across philosophy, religion, and science, a recurring pattern has become visible: the movement from partial understanding to claims of completeness, and from there to forms of absolutism. This pattern is not accidental. It reflects a human tendency to prefer certainty over openness, and closure over tension.

To recognize this tendency is not to eliminate it.

It is to become aware of it.

And that awareness changes something.

It allows us to engage with systems of thought without being fully absorbed by them. It allows us to draw on their insights without accepting their claims to totality. It allows us to remain in a position that is neither fixed nor empty, but responsive.

Human perception does not arrive at a definitive answer.

But we do arrive at a position.

A position that acknowledges that no system fully captures reality, that no perspective is completely neutral, and that no explanation is final. Yet, it affirms that not all interpretations are equally grounded, and that judgment remains necessary.

This stance is not located at one end of a spectrum, but between extremes—between absolutism, which closes the question too quickly, and relativism, which dissolves it entirely. It is a position that accepts tension as a condition rather than a problem to be eliminated.

This does not eliminate the desire for certainty. That desire remains understandable. It reflects the need for stability, for clarity, and for direction in a world that is often complex and unpredictable. But it also reveals a tension: the very conditions that make certainty desirable are the same conditions that make it difficult to achieve without simplification.

In this sense, the question with which we began—why there is something rather than nothing—does not disappear. It remains open. But its role changes. It is no longer something we seek to resolve once and for all. It emerges as part of the context within which we think, act, and understand.

To live without absolutes is not to abandon truth, but to approach it differently. It is to recognize that truth is not possessed in its entirety, but encountered in parts—through perspectives that remain open, through interpretations that remain revisable, and through engagement that does not close what cannot be fully resolved.

Shifts in Understanding

If the previous chapters have shown the limits of absolutism across religion, philosophy, and science, the question remains how this condition has taken shape in more recent history.

We do not stand outside these developments. We are situated within them.

Religions continue to play a central role in addressing existential questions. Even within a world shaped by science and Enlightenment, they have not disappeared. In many regions, they remain strong, and in some cases, they are growing or re-emerging as sources of meaning, identity, and belonging.

Meanwhile, scientific development is accelerating. New discoveries emerge at an unprecedented pace, reshaping our understanding across multiple domains.

In archaeology, for example, findings have challenged earlier assumptions about the development of human civilization. The discovery of Göbekli Tepe [1] in Turkey—dating from approximately 9500–8000 BCE—revealed monumental temple structures built by hunter-gatherers, long before the emergence of agriculture. This suggests that symbolic and religious activity may have preceded settled life, rather than being a consequence of it.

Such findings do not simply add knowledge—they reshape it. What once appeared as a stable narrative becomes open to revision. What seemed foundational can be reconsidered.

Alongside these developments, a profound philosophical shift took place.

Thinkers such as Friedrich Nietzsche articulated a break with the dominance of religious certainty in Western thought. His

1. Schmidt, *Göbekli Tepe.*

declaration: "God is dead. God remains dead. And we have killed him... Is there still any up or down? ... Do we not feel the breath of empty space?" [2] marks not only the rejection of a belief, but the loss of a shared foundation for meaning.

The removal of an absolute framework does not lead directly to clarity. It opens a space in which meaning is no longer given but must be confronted.

Existentialist thinkers explored this condition further, emphasizing freedom, responsibility, and the absence of predetermined meaning. Yet this shift did not produce a universally satisfying alternative. For many, the absence of stable structures has been accompanied by new forms of orientation—consumerism, short-term gratification, and what Nietzsche described as the "last man" a condition marked by comfort, security, and a reduced sense of purpose. This condition does not only reflect comfort and stability; it also introduces the risk of nihilism—the erosion of shared meaning—which had a profound impact on intellectual and cultural developments in the twentieth century.

Later thinkers have continued to reflect on these developments, pointing to patterns of exhaustion, distraction, and the difficulty of sustaining meaning within increasingly complex societies. These analyses do not simply describe decline. They reflect ongoing attempts to understand a changed condition.

At once, alternative perspectives have re-emerged.

The discovery of the Nag Hammadi texts in the twentieth century, including the Gospel of Thomas, revealed that early Christianity was far less unified than later institutional forms suggest. These texts reflect a different emphasis:

"The kingdom is inside you and outside you." [3]

Rather than locating meaning primarily in future fulfillment or external authority, this perspective suggests that understanding is already present, though not always recognized.

2. Nietzsche, *Gay Science*, § 125.

3. Pagels, *Beyond Belief*.

Similar themes appear in various Eastern traditions [4]. In Advaita Vedanta, the distinction between individual self and ultimate reality is questioned. In Buddhism[5], the emphasis lies on awakening to what is already present. In Taoism, balance and unity replace rigid dualism.[6]

These perspectives do not replace earlier traditions. They expand the field of possibilities.

What comes to be visible across these developments is not a single direction, but a widening landscape.

Religious systems continue to exist and evolve. Scientific knowledge continues to expand. Philosophical reflection continues to question and reinterpret. None of these domains has disappeared. None has achieved final dominance.

At the same time, these developments are not confined to abstract thought.

They appear in the world around us.

Religious frameworks continue to shape how individuals and societies understand identity, belonging, and meaning. In many contexts, they remain deeply embedded in cultural and historical narratives, influencing how communities interpret their place in the world.

This influence is not limited to personal belief. It also appears within broader social and geopolitical dynamics. In various regions, religious and ideological frameworks are intertwined with collective identities, historical memory, and political structures. They can provide cohesion and orientation, but also contribute to how differences are perceived, expressed, and, at times, intensified.

What can be observed is that conflicts are rarely defined by a single factor. Yet the ways in which they are understood and experienced are often shaped by underlying frameworks of meaning—religious, cultural, or ideological. These frameworks do not simply explain events after they occur; they can influence how situations are interpreted, how positions are formed, and how

4. Easwaran, *Upanishads*.
5. Rahula, *What the Buddha Taught*.
6. Tao, *Te Ching*.

actions are justified. External perspectives have also challenged dominant assumptions. Figures such as Kandiaronk [7], a Wendat statesman, offered early critiques of European religious and social structures, questioning how societies that claimed moral and religious superiority could however justify inequality, punishment, and the concentration of power. Such perspectives illustrate that what is often presented as universal may, in fact, reflect particular historical and cultural conditions.

Alongside religion, other ideological systems also play a role. They offer narratives through which reality is structured, often providing clarity and direction in complex situations. In doing so, they can strengthen group identity and cohesion but also contribute to divisions when different frameworks encounter one another.

What comes to be visible is not a simple opposition between systems, but a layered interaction. Religious, scientific, and ideological perspectives coexist, overlap, and sometimes reinforce each other, while at other times they stand in tension.

They are part of the broader condition in which human beings continue to search for meaning in a world that is increasingly interconnected yet not unified in how it is understood.

Religions can provide stability and continuity. At the same time, they can contribute to tensions where different interpretations of reality meet, sometimes resulting in divisions between groups that may escalate into conflict.

Alongside religion, ideological and cultural frameworks also influence how people orient themselves. They offer narratives, structure, and direction, particularly in times of uncertainty. These frameworks can strengthen group identity, provide coherence, and guide action.

Within this interplay, narratives and frameworks can also become aligned with positions of authority, shaping how truth is presented and understood, and contributing to the reinforcement of existing structures. This dynamic reflects the fact that science, philosophy, and religion do not only seek to understand truth, but

7. Graeber and Wengrow, *Dawn of Everything.*

participate in shaping how it is interpreted, communicated, and stabilized within society.

We not only seek to understand truth, but also shape how it is perceived, interpreted, and embedded within social and cultural structures.

What can be observed is not the replacement of one system by another, but their coexistence—sometimes reinforcing, sometimes conflicting, often overlapping, and taking on different roles depending on context, interpretation and the structures they operate within.

These dynamics are not isolated. They are part of how human beings continue to seek meaning in a world that is increasingly interconnected and complex.

We are not outside these dynamics.

We participate in them.

This creates a condition in which multiple perspectives remain active meanwhile—each offering insight, each carrying limitations, none providing a final resolution.

The question is no longer which system will ultimately prevail, or which perspective will provide a definitive answer.

It develops into:

How do we navigate between them?

This marks a shift.

From seeking certainty within a single framework to engaging with a plurality of perspectives.

The processes that shape our understanding of truth take place in the mind and respond to different human needs: the search for order and reduction of uncertainty, and the desire to explore, driven by curiosity and the search for new possibilities. Both shape human intention.

Understanding unfolds through movement between openness and stabilization. Without stabilization, there is no clarity. But when stabilization emerges as fixed, it can begin to limit further exploration.

Chapter 6

The Psychology of Absolutism

IF THE PREVIOUS CHAPTERS have shown how systems of thought—philosophy, religion, and science—can move toward absolutism, a further question arises: why does this tendency persist? Why do human beings repeatedly transform partial understanding into claims of completeness?

The answer does not lie only in the systems themselves. It also lies in the structure of human cognition and experience.

What is at stake is not only how systems of thought develop, but why they are needed at all.

The search for truth—whether expressed through religion, science, or philosophy—is not purely abstract. It is rooted in how human beings experience uncertainty, seek orientation, and construct meaning. These motivations are not secondary to understanding; they are part of what makes understanding necessary.

Uncertainty is not merely an intellectual condition; it is felt. To not fully understand introduces a form of instability that is difficult to sustain. Human beings do not only seek truth—they seek orientation. They seek a position from which they can act, decide, and relate to others without constant hesitation.

This creates a natural movement toward certainty and stabilization.

When a system offers clear answers, defined boundaries, and a structured view of reality, it provides more than explanation. It provides relief, but it reduces the need to remain within ambiguity. In this sense, absolutism is not simply an intellectual error. It is also a psychological response.

This response is reinforced by the way in which human cognition operates. Complexity is difficult to maintain. It requires effort, attention, and a willingness to hold unresolved tension. Simpler structures, by contrast, are easier to grasp and communicate. They allow for quicker judgments and more stable conclusions.

What further reinforces this tendency is the way perception itself operates.

Perception is not a passive reflection of reality, but an active construction shaped by prior experience, expectation, and context. Human experience does not encounter reality directly, but through processes of perception and interpretation.

There is no single center within the brain that governs perception, decision-making, or consciousness. Rather, these emerge from the interaction of multiple systems—ranging from higher-order cortical networks to deeper, more automatic processes. What appears unified is, in fact, the result of coordination across distributed structures.

This has important implications.

New information is not encountered in isolation but evaluated in relation to what is already expected. Perception and belief are therefore not independent. They are intertwined.

As a result, there is a tendency to reduce complexity into clearer, more defined forms.

This reduction is not inherently problematic. It is part of how understanding develops. Simplification allows clarity and coordination, but when it becomes fixed, it can also limit further understanding.

Another layer of this movement lies in the role of identity.

Systems of thought are not only frameworks for understanding; they are also frameworks for belonging. To adopt a perspective is often to become part of a group that shares it. Beliefs are embedded in relationships, communities, and shared practices. To question a belief can therefore be experienced not only as intellectual uncertainty, but as a disruption of belonging.

This introduces an additional pressure toward stability.

Positions that are clearly defined are easier to share and defend. Positions that remain open or uncertain are more difficult to maintain socially. Over time, beliefs tend to become more fixed—not only because they are clarified, but because they are reinforced.

This suggests that the tendency toward certainty is not only social but also rooted more deeply in the structure of the individual. In this sense, religion and science do not arise from entirely separate motivations. Both points to responses to the same underlying condition: the need to relate to a reality that is only partially accessible.

They differ in method, language, and scope. Yet both seek to reduce uncertainty, establish orientation, and provide a framework within which human experience can be understood.

From this perspective, religion can also be understood as operating within these psychological dynamics.

It provides structure, reduces uncertainty, and strengthens social cohesion. Research indicates [1] that religious engagement can reduce stress, anxiety, and feelings of isolation. It offers narratives that organize experience and give direction.

At the same time, the same mechanisms can reinforce boundaries, strengthen group distinctions, and contribute to the persistence of beliefs that resist revision. Religion, in this sense, reflects both the adaptive and the limiting aspects of human cognition.

Beneath conscious reasoning, there are layers of motivation, conflict, and interpretation that are not always directly accessible. Thinkers such as Sigmund Freud [2] brought this into focus by introducing the concept of the unconscious. In this view, human behavior is not fully governed by conscious intention but shaped by underlying drives and unresolved tensions.

From this perspective, absolutism can function as a way of stabilizing internal conflict.

1. Haidt, *Righteous Mind.*
2. Freud, *Ego and the Id.*

A different, though related, perspective can be found in Carl Jung [3], who emphasized symbolic structures and recurring patterns within the psyche. Rather than focusing only on repression, he pointed to deeper frameworks through which meaning is organized. In this sense, his view resonates with philosophical traditions such as those associated with Plato [4], where underlying structures shape perception.

Both perspectives point to the same conclusion: understanding is not purely rational.

This does not mean that reason is unimportant. It means that reason does not operate alone. What individuals take to be clear, necessary, or convincing is shaped not only by argument, but also by inner tension, memory, expectation, and the need for coherence. The structure of belief is therefore never purely logical.

This is especially evident in the experience of anxiety. Søren Kierkegaard [5] described anxiety as arising not from external threat, but from freedom itself. The ability to choose introduces uncertainty. To act means to commit without complete knowledge. This creates a tension that is intrinsic to human existence.

In such conditions, frameworks that reduce ambiguity become more appealing.

There are moments in which this comes to be particularly visible.

You do not experience your life only from within but also observe it as if from a distance. You act, speak, and respond, while meanwhile recognizing that what unfolds does not align with what you intend. You see your own role in it, and yet you are unable to fully redirect the situation.

This experience is not limited to a single domain. It appears in relationships, in work, and in everyday interactions. You recognize patterns that repeat, even as you attempt to change them. Effort does not always resolve them; it can even intensify them. What

3. Jung, *Archetypes and the Collective Unconscious.*
4. Plato, *Republic.*
5. Kierkegaard, *Fear and Trembling.*

takes shape as visible is not simply error, but a more fundamental tension between intention, action, and outcome.

In such moments, the distinction between control and limitation becomes less clear. You neither fully determine the situation nor stand outside it—you are part of it. Understanding begins within this condition.

It doesn't represent a failure of reasoning, but a condition in which understanding operates. The difficulty of maintaining openness is not only conceptual. It is grounded in the way experience itself unfolds.

Modern psychology further clarifies this dynamic. The concept of cognitive dissonance [6] describes the discomfort that arises when beliefs and actions are not aligned. Rather than maintaining this tension, individuals tend to resolve it—often by adjusting beliefs to restore coherence. What begins as inconsistency is transformed into certainty.

This becomes more evident when considering how beliefs are maintained.

Human cognition does not process information in a purely neutral way. It filters, selects, and reinforces patterns that are already present. Expectations influence perception, and perception reinforces expectations.

Cognitive dissonance illustrates this clearly: when new information conflicts with existing beliefs, the discomfort that arises is often resolved not by changing the belief, but by reinterpreting the information.

Unconscious biases further shape this development. Emotional responses, prior experiences, and subtle influences—such as those present in media and social interaction—can guide perception without entering conscious awareness. Over time, these processes can form reinforcing cycles in which beliefs become increasingly stable, even when confronted with contradiction.

At the same time, social norms reinforce this movement. Beliefs that align with a group provide stability and belonging.

6. Festinger, *Theory of Cognitive Dissonance.*

Beliefs that deviate introduce tension—not only intellectually, but relationally.

These layers—cognitive, emotional, and social—already explain much of the tendency toward absolutism. Yet recent developments in neuroscience introduce an additional dimension. [7]

What we experience as reality is not a direct representation of the world, but a constructed interpretation.

The brain actively organizes sensory input into coherent forms that allow for action. This construction is functional—it enables effective interaction with the environment—but it is not necessarily complete.

Different organisms perceive the world differently yet function successfully within it. This suggests that perception does not aim at full accuracy, but at usefulness.

From this perspective, what we experience as "reality" is shaped by the conditions under which perception operates. The observer is not separate from the process of understanding, but part of it.

The brain does not simply receive information—it actively organizes, filters, and interprets it. What appears as immediate and obvious is already shaped by underlying processes.

Some contemporary perspectives extend this even further.

Thinkers such as Annaka Harris [8] have raised the possibility that consciousness may not simply emerge from complex physical systems but could be a more fundamental aspect of reality itself. In this view, consciousness is not fully reducible to neural processes but may be present in more basic forms across different levels of existence.

This perspective remains speculative and is not widely established within scientific consensus. Yet it introduces an important shift. It suggests that the relationship between mind and reality may be more complex than a purely reductionist account allows.

What is significant here is not that a new definitive answer is provided, but that the boundaries of scientific inquiry are

7. Damasio and Damasio, *Self Comes Into Mind.*

8. Harris, *Conscious.*

expanding. This raises a fundamental question: how much of what we take to be reality is dependent on how it is constructed?

Consider perception itself. Different organisms experience the world in fundamentally different ways. An octopus, for example, processes visual information through a nervous system that differs significantly from that of humans. Its perception of color, texture, and environment is not identical to ours. And yet, both humans and octopuses interact effectively with their environments.

Humans respond to traffic signals without difficulty. Octopuses camouflage themselves with remarkable precision. In both cases, behavior appears coherent, even though the underlying experience of reality may differ substantially.

This suggests that what we take to be "reality" is not a single, fully shared experience, but a functional construction that allows organisms to navigate their environment.

What matters here is not that reality comes to be unreal, but that access to it is mediated. Different forms of life do not require identical experience in order to function effectively within the world. This reinforces the idea that what appears immediate and self-evident may already be shaped by the specific conditions under which it is perceived.

A similar tension appears in the debate on free will.

Thinkers such as Daniel Dennett [9] argue for a compatibilist position, suggesting that free will can be understood within a structured, causal framework. In this view, human agency is not independent of causation but operates within it in a meaningful way.

In contrast, perspectives associated with figures such as Robert Sapolsky [10], Dick Swaab [11], and earlier thinkers such as Baruch Spinoza [12] emphasize determinism. From this perspective, human behavior is the result of prior causes—biological, environmental,

9. Dennett, *Freedom Evolves.*
10. Sapolsky, *Determined.*
11. Swaab, *We Are Our Brains.*
12. Spinoza, *Ethics.*

and neurological. What is experienced as choice may be better understood as the outcome of processes that are not freely initiated.

The implications of this are significant. If free will is limited or absent, then the notion of an autonomous self becomes less stable. If it is retained, even in a limited sense, then the self remains a meaningful point of reference.

What is striking is not only the difference between these positions, but the pattern they reveal.

Even within empirical science, interpretation plays a role. The same body of evidence can support different conceptual models. One emphasizes structure and causation; another emphasizes agency and meaning. The tension between these perspectives mirrors the broader pattern observed throughout this book: partial understanding extended toward more comprehensive claims.

At this point, the emergence of absolutism comes to be more understandable.

It is supported by multiple layers: the need for cognitive simplicity, the desire for social belonging, the management of internal tension, the constructed nature of perception, and the ambiguity of agency itself.

To move beyond absolutism, therefore, requires more than intellectual critique.

It requires awareness of the conditions that make it appealing.

To recognize that what appears certain may be constructed, what feels stable may be reinforced, and what seems complete may remain partial. This does not eliminate the need for understanding.

But it changes how we relate to it.

If understanding is shaped within the mind, the question persists how it is expressed—and how language gives form to what we think we know.

If perception and interpretation shape how reality is experienced, language shapes how that experience becomes thinkable and shared.

Chapter 7

Language, Interpretation, and the Limits of Meaning

IF THE PREVIOUS CHAPTER has shown that understanding is shaped not only by cognition but also by deeper psychological and perceptual structures, a further dimension must be considered: the role of language.

Language is not only the medium through which understanding is expressed—it is the means through which existential questions themselves are transmitted, shared, and reshaped across individuals and generations.

It is through language that understanding becomes expressible, shareable, and stable. Without language, thought remains largely internal and transient. With language, it develops into structured, communicated, and preserved. Language allows systems of thought to emerge, develop, and endure across time.

Language does not only carry meaning—it shapes the conditions under which meaning can appear.

Yet this same function introduces a limitation.

Meaning is shaped through interaction between people.

This shaping function often remains unnoticed because it operates continuously. Words do not appear as filters, but as neutral carriers of meaning. Yet the structure of language influences what can be expressed clearly, what remains implicit, and what comes to be difficult to articulate.

In this sense, language does not only communicate thought—it conditions it.

Language as Representation

This question of mediation is not new. René Descartes [1] already introduced a fundamental distinction between the external world and the ideas through which it is represented in the mind. What is perceived is not the world itself, but the representation of it within the mind. This raises a central question: to what extent do these representations correspond to reality as it is? The possibility that they may not fully do so introduces a gap between experience and truth.

To understand this more clearly, it is important to recognize that language does not operate on a neutral foundation. Understanding is already structured before it is expressed. Thinkers such as Arthur Schopenhauer [2] argued that what we experience as reality is not a direct encounter with the world as it is, but a form of representation shaped by perception and cognition. What appears immediate is already interpreted.

This introduces a layering effect. Reality is not encountered directly, but through perception, and that perception is already structured before it is expressed in language. By the time something is expressed in language, it has already passed through multiple stages of interpretation.

Language does not create this structure, but it builds upon it. What is expressed is therefore never a direct reflection of reality, but a further step within an already mediated process.

At the same time, human cognition does not aim to represent reality in its full complexity. It reduces, filters, and organizes information in ways that allow for action. What is perceived is not

1. Descartes, *Meditations on First Philosophy.*
2. Schopenhauer, *World as Will and Representation.*

everything that exists, but what is relevant within a limited capacity for attention and decision-making. This simplification is not a flaw—it is necessary for functioning. Without it, action would be overwhelmed by detail.

This reduction is therefore functional. It allows attention to be directed, decisions to be made, and interaction to take place without being absorbed by the full complexity of what exists. What is left out is not irrelevant but temporarily set aside in order to make action possible.

Yet this necessity introduces a gap.

What is experienced is already a reduction. What is expressed through language is a further reduction. What begins as perception gradually stabilizes into interpretation, and over time into belief.

Something is always left out of this process.

What is left out does not disappear. It remains present as a background that is no longer explicitly recognized. This creates the possibility that what is excluded from expression may still influence how meaning is understood.

In everyday situations, this gap often remains unnoticed. Communication functions well enough for practical purposes. Instructions are followed, agreements are made, and shared understanding appears stable. Yet when language is used to describe more complex or abstract domains—such as morality, existence, or reality itself—the limitations become more visible.

Concepts that appear clear at first may reveal multiple interpretations upon closer examination.

Terms such as "truth," "good," or "self" suggest stability, yet their meaning varies across contexts, traditions, and individuals. What one system defines as truth may differ from another. What one framework presents as self-evident may, from another perspective, appear constructed.

The apparent clarity of language conceals the variability of interpretation.

Clarity, in this sense, can be misleading. What appears precise may depend on shared assumptions that are not explicitly stated.

When those assumptions differ, the same words can carry different meanings without this being immediately recognized.

This variability is not inherently problematic. It reflects the richness of human understanding. However, it comes to be problematic when the flexibility of language is no longer recognized.

When a particular meaning is fixed and treated as the only valid interpretation, language begins to function as a boundary rather than a medium. Words no longer point toward reality; they define it.

At this point, language begins to function less as a medium of exploration and more as a framework of stabilization. Instead of opening possibilities, it organizes them into fixed structures that are treated as given.

This is one of the mechanisms through which absolutism begins to take shape.

Once a term is stabilized within a system, it emerges as part of a structured framework. It is repeated, taught, and used consistently within a community. Over time, this usage appears natural. The interpretation emerges as embedded in the language itself, and alternative meanings become less visible.

At this point, questioning the meaning of a term can appear as questioning reality itself.

Language and the Limits of Expression

The role of language in shaping how reality is understood becomes particularly visible in modern philosophy. As Ludwig Wittgenstein[3] famously stated, "The limits of my language mean the limits of my world." This formulation points to a radical implication: what can be expressed is not separate from what can be understood. It suggests that meaning emerges within the context of language itself. Language does not merely describe reality—it shapes the horizon within which reality appears. Language, in this sense, does

3. Wittgenstein, *Tractatus Logico-Philosophicus.*

not stand between us and reality—it defines the horizon within which reality becomes intelligible.

This insight has important implications for how truth is approached. If understanding is always mediated through language, then no description of reality can be entirely independent of the forms through which it is expressed. What appears as a statement about reality is also shaped by the conceptual and linguistic framework within which it is formulated.

In his later work, Ludwig Wittgenstein[4] shifted this perspective. Rather than viewing language primarily as a structure that reflects the limits of meaning, he emphasized its practical use within specific social contexts. Meaning, in this view, is not determined by correspondence to reality, but by how words are used within forms of life. Language functions less as a mirror of the world and more as a tool embedded in human activity.

This shift introduces a different dimension to the role of language. Understanding is not only constrained by what can be expressed, but also shaped by how language is used in practice. Meaning becomes dynamic, context-dependent, and tied to interaction rather than fixed representation.

Taken together, these perspectives suggest that meaning is neither a direct reflection of reality nor entirely independent of it, but emerges within the interplay of perception, language, and use.

From this perspective, language can be approached in different ways. In one sense, it marks the limits of what can be meaningfully said. In another, it functions as a tool through which claims about reality are formulated, tested, and revised, as emphasized by Karl Popper [5]. In yet another sense, it reflects the way reality is represented through human perception, as suggested by Arthur Schopenhauer, for whom the world appears always as representation.

These perspectives do not exclude one another. They reveal different dimensions of how language mediates the relation between human understanding and reality.

4. Wittgenstein, *Philosophical Investigations.*

5. Popper, *Logic of Scientific Discovery.*

Language, Testing, and the Openness of Truth

A similar openness is reflected in the work of Karl Popper [6], who argued that knowledge does not advance through final confirmation, but through continuous testing and revision. Scientific theories are not established as definitive truths, but remain provisional, open to being challenged and refined. This introduces a dynamic view of truth—not as something possessed, but as something approached through an ongoing process of correction and adjustment.

Paradigms and Collective Frameworks

This mediation is not only individual— it is also collective. Thomas Kuhn [7] introduced the concept of paradigms to describe how shared frameworks shape what is recognized as knowledge within a given period. What counts as evidence, what questions are asked, and what explanations are accepted are all influenced by the linguistic and conceptual structures within which a community operates.

When such frameworks shift, the understanding of reality itself can change—not because reality has altered, but because the way it is interpreted and articulated has been transformed.

What is recognized as 'truth' is therefore not only a matter of evidence, but also of the framework within which that evidence is interpreted.

Language and Religious Interpretation

If language shapes what can be expressed and understood, then religious texts are not only sources of meaning, but also products of interpretation, transmission, and linguistic framing.

This dynamic is particularly evident in the interpretation of foundational texts. Religious traditions rely on language to

6. Popper, *Conjectures and Refutations.*
7. Kuhn, *Structure of Scientific Revolutions.*

preserve and transmit their teachings. Yet these texts are not static in their meaning. They are interpreted within historical, cultural, and linguistic contexts.

Further methods developed within historical scholarship illustrate how interpretation operates and has shown that many biblical texts underwent processes of transmission, variation, and modification. Differences in wording, emphasis, and interpretation emerged over time, reflecting the perspectives and concerns of different communities.

This illustrates that meaning is not static, even when it is presented as such. What appears fixed is often the result of processes that have stabilized interpretation over time. The stability of meaning reflects continuity of use, not necessarily completeness of understanding.

What is often treated as a fixed and unified message is, in many cases, the result of layered interpretation.

Bart Ehrman has described criteria used to assess early texts, including the criterion of dissimilarity and the criterion of embarrassment. [8]

The criterion of dissimilarity suggests that sayings or events that differ from both earlier traditions and later doctrinal developments may be less likely to have been shaped by those traditions and therefore may reflect an earlier layer of interpretation. The criterion of embarrassment, by contrast, highlights elements that would have been difficult or inconvenient for the communities that transmitted them. Such elements are considered less likely to have been invented, precisely because they do not easily support a coherent or favorable narrative.

These criteria do not produce certainty. They illustrate the opposite. What appears as a fixed account is approached through methods that attempt to distinguish layers of interpretation, reconstruction, and transmission. Meaning is not simply given—it is inferred, evaluated, and continuously reinterpreted.

8. Ehrman, *Jesus: Apocalyptic Prophet of the New Millennium.*

This does not diminish the significance of these texts. But it highlights that meaning is not simply contained within them—it is constructed through engagement with them.

A similar process occurs outside of religious contexts. Scientific language, while more formally defined, also involves interpretation. Terms such as "gene," "information," or "cause" operate within conceptual frameworks that shape how they are understood. As these frameworks evolve, so too does the meaning of the terms.

Even in philosophy, where precision is emphasized, language remains both a tool and a constraint. Concepts are defined in order to reduce ambiguity, but those definitions can themselves become fixed, limiting alternative ways of thinking.

This reflects a broader pattern.

Language enables understanding, but it also stabilizes it.

This dual function is essential. Without stabilization, communication would not be possible. Without flexibility, understanding would not develop. The tension between these two aspects is not something that can be resolved, but something that must be managed.

What begins as an attempt to describe reality develops into, over time, a structure that organizes perception. The words we use influence not only how we communicate, but how we interpret what we encounter.

This influence emerges as particularly significant in contexts where communication is accelerated.

When meaning must be conveyed quickly and efficiently, there is a tendency to favor clarity over nuance. Complex ideas are reduced to simpler forms. This reduction often takes place incrementally. Each step simplifies slightly, but the cumulative effect can be substantial. Over time, what was once a nuanced position may be represented by a more rigid and simplified expression. Ambiguity is minimized. Positions become more defined, more distinct, and more easily recognizable.

This dynamic makes communication more efficient, but it also narrows interpretation.

Nuanced positions, which require time and context to fully understand, become more difficult to sustain. In their place, clearer and more definitive statements emerge—not necessarily because they are more accurate, but because they are easier to communicate and repeat.

Over time, this reinforces a movement toward simplification.

And with simplification comes a greater risk of absolutism. Simplified expressions are not only easier to understand—they are also easier to repeat. What can be expressed clearly and quickly is more likely to circulate, to be recognized, and to become established. In this way, language does not only reflect understanding; it influences which forms of understanding persist.

When language reduces complexity into fixed expressions, and those expressions are repeated within stable frameworks, the space for interpretation contracts. Alternative meanings are less frequently considered. The flexibility that originally allowed language to adapt becomes constrained.

Yet language does not operate in isolation.

What is expressed in words does not move unchanged from one individual to another. It is interpreted, reshaped, and recontextualized within interaction. Communication is not a linear process in which meaning is transferred intact. It is a dynamic process in which meaning is continuously reconstructed.

Once meaning is expressed through language, it no longer remains confined to the individual. It circulates, is repeated, and comes to be part of a shared environment. Within this collective dimension, interpretations are reinforced, simplified, and sometimes polarized.

What is said is only part of what is understood. Interpretation depends on context, prior experience, expectations, and the values that shape how a message is received. The distinction between sender and receiver is therefore less fixed than it appears. Within interaction, roles shift, and meaning emerges between participants rather than residing solely in what is expressed.

In this sense, communication does not simply convey meaning. It participates in its formation.

The quality of this process depends on multiple factors. Information that is incomplete, delayed, or lacking context does not merely reduce clarity—it reshapes interpretation. What is available becomes what is taken to be real within a given situation.

At the same time, communication takes place within cultural frameworks. Norms, values, and expectations influence which perspectives are expressed, which are reinforced, and which remain unspoken. Over time, these patterns contribute to the stabilization of shared interpretations.

This adds an extra layer to the process already described. Meaning is not only reduced through language. It is further shaped through interaction, reinforced through repetition, and stabilized within social contexts.

A further development extends this movement.

Systems capable of generating language—based on patterns derived from large collections of human communication—are increasingly involved in how meaning is produced and circulated. These systems do not possess understanding in the human sense, yet they generate coherent expressions that participate in the same processes of interpretation and reinforcement.

What appears as new content is often a recombination of existing patterns. In doing so, such systems can reflect and amplify prevailing structures of meaning.

This raises a similar question to those encountered earlier: to what extent does communication clarify understanding, and to what extent does it reinforce existing patterns?

The role of language is no longer limited to human interaction. It develops into part of a broader system in which meaning is produced, circulated, and stabilized at scale.

The influence of communication on the shaping of meaning is not new, but its form has changed over time.

Earlier forms such as advertising and propaganda already demonstrated how language and framing could influence perception. By selecting certain aspects of reality and presenting them in specific ways, they shaped how situations were understood, often without this influence being fully recognized.

With the development of mass media, and later social media, this process expanded and accelerated. Information could be distributed more widely and more rapidly. At once, the structures that filtered and verified information became less centralized. This allowed for a greater diversity of perspectives, but also increased the circulation of simplified, selective, or misleading interpretations. What gains attention is not always what is most complete, but what is most easily communicated and shared.

These developments illustrate how communication does not merely transmit ideas—it shapes which ideas persist, and how they are understood.

A further shift is now taking place.

Interactive systems capable of generating and responding to language allow individuals to engage directly with structured information in ways that were not previously possible. Questions can be explored, perspectives compared, and interpretations tested in a more dynamic and iterative manner.

This introduces a different relationship to meaning.

Where earlier forms of communication primarily broadcast or distributed interpretations, these systems enable interaction with them. This does not eliminate the limitations already described. The underlying patterns, assumptions, and structures remain present. But the way individuals can engage with them comes to be more flexible.

What remains consistent across these developments is the role of communication as a shaping force.

It does not simply reflect reality.

It participates in how reality is understood.

This does not mean that language should be abandoned as a tool for understanding. It is indispensable. But it must be approached with an awareness of its limits.

To engage with language critically is not to reject it, but to recognize that words do not fully capture what they refer to.

They point.

They indicate.

But they do not exhaust.

This awareness introduces a different way of relating to meaning.

Instead of treating language as a final representation of reality, it can be understood as a provisional framework—one that allows us to engage with complexity but does not eliminate it.

In this sense, interpretation remains open.

Different perspectives are not necessarily errors to be corrected, but possibilities to be examined. The goal is not to fix meaning once and for all, but to refine understanding while remaining aware of its limits.

To remain aware of this doesn't require rejecting language but engaging with it more carefully. It involves recognizing when clarity conceals complexity, when definitions become fixed, and when interpretation narrows without being noticed.

This position does not provide the clarity of fixed definitions, nor the simplicity of absolute statements.

But it preserves something essential:

the capacity to reconsider.

Across these perspectives, language does not simply communicate truth—it participates in shaping what can be recognized, tested, and accepted as truth.

If human communication shapes how meaning is expressed and stabilized, then the broader structures operates within which language—social, cultural, and communicative—play a decisive role in how understanding develops.

Language, Artificial Intelligence, and the Production of Meaning

The role of language has taken on a new dimension in recent developments in artificial intelligence. Systems based on large-scale language processing operate entirely through patterns derived from human communication. In these systems, language is not only a medium of expression—it becomes the primary structure through which knowledge is represented, generated, and circulated.

Human knowledge, cultural patterns, and abstract reasoning are extensively encoded in language. As a result, such systems can generate coherent and contextually appropriate responses that resemble understanding. Yet this raises a fundamental question: to what extent does linguistic coherence correspond to understanding or truth?

Language, as explored throughout this chapter, is not neutral. It involves selection, simplification, and contextual framing. These characteristics are not removed in computational systems; they are amplified. What appears as a structured response may reflect patterns of language use rather than direct engagement with reality.

This introduces a tension. If meaning is generated through language, and language itself is shaped by interpretation, then outputs based purely on linguistic patterns may produce meaning without grounding in lived experience, perception, or direct interaction with the world.

From the perspective of Karl Popper [9], such outputs can be approached as hypotheses rather than established knowledge—provisional formulations that require testing, comparison, and critical evaluation. Without such processes, there is a risk that linguistic coherence is mistaken for validity.

At the same moment, this development reflects insights associated with Ludwig Wittgenstein. [10] Meaning emerges through use, context, and interaction. Systems that operate through large-scale linguistic exchange participate in this dynamic, extending it beyond individual human interaction into broader, distributed processes.

This raises a further question. If language-based systems increasingly shape how information is accessed and interpreted, how can the openness of interpretation be preserved? How can language remain a medium for exploration rather than becoming a source of fixed and unexamined outputs?

The challenge is not the presence of such systems, but how they are engaged. When treated as authoritative sources, they

9. Popper, *Conjectures and Refutations.*

10. Wittgenstein, *Philosophical Investigations.*

risk reinforcing existing patterns and stabilizing particular interpretations. When approached critically, they can function as tools that support exploration, comparison, and refinement of understanding.

This development does not fundamentally alter the role of language described in this chapter. It intensifies it. Language continues to enable understanding, but also shapes, filters, and stabilizes it. What changes is the scale, speed, and influence with which these processes occur.

Chapter 8

Society, Communication, and the Formation of Polarization

IF UNDERSTANDING IS SHAPED by cognition, reinforced by psychological needs, and stabilized through language, then the broader structures within which communication takes place play a decisive role in how perspectives develop.

Human beings do not think in isolation. They think within environments—social, cultural, and communicative—that influence how ideas are formed, expressed, and sustained. These environments do not determine what individuals believe, but they shape the conditions under which beliefs become more or less likely.

Across history, shared narratives have played a central role in shaping these environments. Large groups of people are able to cooperate not only through biological or material connections, but through shared ideas, stories, and symbols. These shared frameworks provide coherence and enable coordination across individuals who do not know each other personally.

Early symbolic artifacts, such as those discussed by Neil MacGregor [1], suggest that the human capacity to represent and reinterpret reality through symbols has deep historical roots. The Lion Man, a figure dating back more than 40,000 years, illustrates

1. MacGregor, *History of the World.*

this early symbolic imagination. In the present, similar symbolic acts continue to shape meaning—for example, the restoration of religious structures in historically charged locations. Such acts demonstrate the enduring role of cultural symbols in organizing collective identity.

At the same time, these same narratives can function as markers of distinction. What unites one group can differentiate it from another. In this way, the structures that enable cooperation also carry the potential for division.

This influence is often indirect, shaping which perspectives become visible, accessible, and sustainable. It does not impose specific conclusions. In this way, the environment participates in shaping understanding without fully determining it.

One of the most significant features of such environments is the way communication is structured.

In contexts where communication is slow, extended, and reflective, complexity can be maintained more easily. Ideas can be developed gradually, distinctions can be clarified, and ambiguity can be explored without immediate pressure to resolve it. Understanding in such contexts often remain open-ended.

This openness allows for hesitation, reconsideration, and the gradual refinement of ideas. Positions are not immediately fixed, and differences can remain without requiring immediate resolution. The pace of communication supports a broader range of interpretation.

In contrast, in environments where communication is rapid and continuous, different patterns emerge.

When responses are expected quickly, there is less space for reflection. Positions must be formulated in shorter time frames, often without the opportunity to fully consider alternatives. Under these conditions, clarity becomes more valuable than nuance. Statements that are easily understood and quickly recognizable are more likely to be repeated and retained.

Clarity, in this context, functions as a form of efficiency. It allows communication to proceed quickly and reduces the effort required to interpret meaning. Yet this efficiency often comes at the cost of detail. What is not immediately clear is more easily overlooked.

Not really this does necessarily lead to error, but it does lead to simplification.

Simplification can be useful. It enables shared understanding and coordinates action. However, when it becomes the dominant mode of communication, it begins to limit the range of perspectives that can be expressed and maintained.

Positions that present themselves with clarity and certainty are easier to communicate. They require less explanation, leave less room for ambiguity, and are more easily shared within groups. In this sense, absolutism is not only a philosophical position—it is also a communicative advantage.

Yet what is easy to communicate is not always what is most accurate. The reduction of complexity into clear and stable forms makes ideas more effective, but also more limited. What is gained in clarity may be lost in depth.

Over time, this dynamic reinforces the preference for certainty over nuance, not because it is more accurate, but because it is more easily sustained within shared communication.

This reinforcement does not require conscious intention. It emerges through repeated interaction. What is expressed more frequently becomes more familiar, and what is familiar comes to be more readily accepted.

Complex positions are reduced to more defined forms. Ambiguity is minimized. Ideas that might otherwise remain open are presented in ways that appear more complete. Over time, this creates a tendency toward more stable and often bounded perspectives.

This tendency is reinforced by the social dimension of communication.

Individuals do not only express ideas; they respond to others. Communication is relational. It involves agreement, disagreement, alignment, and differentiation. Within this development positions are not only clarified—they are also contrasted.

To understand one perspective often involves distinguishing it from another.

This dynamic of distinction is not neutral. It often highlights differences more often than similarities. What overlaps becomes less visible, while what contrasts comes to be more pronounced.

As this process intensifies, distinctions can become sharper. Differences that were once gradual or overlapping begin to appear more discrete. Positions become more clearly separated, and the space between them emerges as less visible.

This is one of the pathways through which polarization emerges.

Thinkers such as Yuval Noah Harari [2] have emphasized the role of shared narratives in enabling cooperation at scale. Similar narrative structures appear across other influential interpretations of history, where complex developments are organized into coherent and communicable frameworks.

Alternative perspectives highlight the contingency and fragmentation of historical processes, suggesting that such narratives may simplify developments that are less linear and more context-dependent than they appear.

Historical developments illustrate that such processes are rarely linear or driven by a single factor. Interpretations that attribute large-scale developments to a single factor often simplify a more complex reality. Cultural, political, environmental, and biological factors interact in ways that are difficult to disentangle.

Different perspectives on history reflect this complexity. Some emphasize the role of shared narratives and belief systems in enabling cooperation at scale, while others point to the contingent and often chaotic nature of historical development, in which multiple factors converge.

What comes to be visible is not a single explanatory pattern, but a field of interacting influences. The same tendency toward simplification that shapes communication in the present also appears in how history is interpreted. This tension between

2. Harari, *Sapiens.*

cooperation and conflict can also be approached from a broader perspective.

Friedrich Nietzsche [3] argued that life itself is not primarily oriented toward harmony, but toward expansion, struggle, and expression of force. What appears as conflict at the human level may reflect more fundamental dynamics that are not unique to human societies.

In the natural world, living organisms compete, adapt, and expand within their environments. At the same time, forms of cooperation exist among animals, often structured through learned behavior and social organization. Certain species display patterns that resemble cultural transmission, such as shared hunting strategies or group-specific behaviors.

What distinguishes human societies is not the absence of these dynamics, but the way they are expressed and interpreted. Human beings do not only act within these patterns—they interpret them. Through language, symbols, and shared narratives, conflict and cooperation become embedded in systems of meaning. Differences are not only experienced, but explained, justified, and sustained within conceptual frameworks.

In this transition from nature to culture, the dynamics of interaction become layered with interpretation. What might otherwise remain situational becomes structured and enduring.

This introduces a tension that remains present. Human societies often present themselves as structured, rational, and guided by shared values, while the dynamics that underlie conflict may not be entirely separate from more fundamental patterns of interaction found in the natural world—competition, expansion, and the negotiation of space and influence.

Through language, symbols, and systems of meaning, conflict takes shape as framed, justified, and organized. It is no longer only enacted—it is interpreted.

This creates the possibility that what appears as purely ideological disagreement may also reflect underlying patterns that are less visible, but no less active.

3. Nietzsche, *Gay Science*.

To recognize this doesn't resolve the tension.

But it makes it visible.

Polarization, in this sense, develops gradually. It does not begin with fully opposed positions, but with small distinctions that become more defined over time. As these distinctions are repeated and reinforced, they begin to structure how perspectives are understood.

Polarization is not simply the result of disagreement, but of how disagreement is structured and reinforced. It takes shape as polarization when perspectives are increasingly defined in opposition to one another, and when the complexity within each perspective is reduced.

At this point, positions begin to function less as attempts to understand reality and more as markers of alignment.

This shift is reinforced by repetition.

When certain expressions are used consistently within a group, they become familiar and recognizable. Familiarity creates stability. Over time, these expressions can become shorthand for more complex ideas. Instead of engaging with the full depth of a perspective, individuals engage with its simplified representation.

What is gained in efficiency is lost in depth. The original complexity of a perspective is condensed into a form that is easier to recognize, but less capable of capturing its full meaning.

This makes communication more efficient, but it also reduces nuance.

Once simplified representations become dominant, alternative interpretations become less visible. The range of possible meanings narrows, not necessarily because they no longer exist, but because they are less frequently expressed and less easily recognized.

This dynamic is further amplified by feedback loops, particularly in contemporary digital environments.

In contemporary contexts, these feedback mechanisms are intensified by digital communication environments. The speed and scale at which information circulates allow for rapid reinforcement of perspectives. Content that is clear, emotionally engaging, or easily recognizable is more likely to be shared and repeated.

And yet, the structures through which information is distributed can influence which perspectives are encountered more frequently. This does not eliminate exposure to alternative views, but it can shape the balance of what is visible and what remains peripheral.

Emerging technologies introduce an additional layer. Systems capable of generating and organizing information at scale participate in the circulation of meaning. They do not determine what is true, but they influence how information is accessed, compared, and interpreted. This creates new possibilities for exploration, but also new forms of reinforcement.

Feedback loops do not only reinforce agreement; they also reduce exposure to alternative perspectives. When interaction occurs primarily within aligned contexts, the range of considered viewpoints narrows without necessarily being recognized as such.

When individuals encounter perspectives that align with their own, those perspectives are reinforced. Agreement confirms the stability of a position. Even so, encountering opposing views can lead to further clarification and strengthening of one's own perspective—not necessarily through examination, but through contrast.

Over time, this can lead to a situation in which perspectives become more internally consistent, but less responsive to alternatives.

It doesn't necessarily require deliberate exclusion. The process can unfold without explicit intention. Individuals may not seek to exclude other perspectives, yet the structure of interaction can still lead to their gradual disappearance from view. It can emerge naturally from interaction itself.

At this stage, the role of language comes to be particularly significant. As discussed in the previous chapter, language stabilizes meaning. In a polarized environment, this stabilization develops into more rigid. Terms are used consistently within a perspective, and their meaning takes shape as increasingly fixed.

At the same time, those same terms may carry different meanings in different contexts.

This creates a situation in which communication across perspectives becomes more difficult—not only because of disagreement, but because the same words no longer refer to the same concepts. What appears as contradiction may, in part, be a difference in interpretation.

Recognizing this possibility requires a shift in attention—from what is being said to how it is being understood. Without this shift, differences in meaning are easily interpreted as differences in truth.

Yet this difference is not always recognized.

Instead, it can be interpreted as error, distortion, or opposition. The possibility that meaning itself may differ comes to be less visible. The focus shifts from understanding to correction.

This reinforces the boundary between perspectives.

Once boundaries are established, they tend to be maintained.

Positions that align with a framework are integrated and reinforced. Positions that challenge it are more likely to be dismissed or reinterpreted. Over time, the framework takes shape as more stable, but also more closed.

Stability, in this sense, is achieved through reduction. What does not fit is less frequently considered, and over time, may no longer appear as a viable alternative.

This pattern reflects the same structure observed in earlier chapters.

What begins as an attempt to understand develops into, through stabilization and reinforcement, a system that appears complete. The conditions that initially allowed for exploration become conditions that limit it.

At the collective level, absolutism emerges not only through belief, but through interaction.

It is not imposed from a single source. It develops through repeated processes of simplification, alignment, and reinforcement.

Each of these processes is individually understandable. Together, they create a structure that becomes increasingly resistant to change. What began as a flexible perspective can, through repetition, take on the appearance of necessity.

It doesn't imply that communication inevitably leads to polarization. The same structures that enable simplification can also support complexity, depending on how they are engaged. Environments that allow for slower reflection, sustained dialogue, and openness to revision can maintain a broader range of perspectives.

The question, therefore, is not whether communication shapes understanding—it always does—but how it does so.

To recognize these patterns is not to stand outside them, but to engage with them more consciously.

Such awareness does not eliminate disagreement, nor does it resolve differences.

It changes the way disagreement is interpreted. Instead of indicating that one position must be entirely correct and the other entirely mistaken, disagreement can be seen as a difference in how reality is being approached and structured.

But it changes how perspectives are approached.

Instead of treating opposing perspectives as fixed and complete, they can be understood as outcomes of processes that are themselves dynamic and open to examination. The focus shifts from defending positions to understanding how those positions comes to be.

It doesn't weaken conviction.

It situates it.

Throughout history, different systems of organization—whether philosophical, political, or religious—have presented themselves as complete frameworks. While differing in content, they often share similar structural tendencies toward coherence, stability, and exclusion of alternatives.

Systems of information that aim to present reality objectively operate within the same constraints of framing, selection, and interpretation. Complete neutrality is difficult to maintain, as perspective is inherent in the act of representation.

To recognize these patterns is not to stand outside them. It is to engage with them more consciously. Communication continues, perspectives continue to form, and differences continue to emerge. But within this, there remains the possibility of maintaining awareness of how these processes unfold.

If perspectives are shaped through layered processes—formed by cognition, stabilized through language, and reinforced within social and communicative environments—then the question arises how individuals can engage with these conditions more consciously.

The patterns described are not abstract. They are visible in how ideas circulate, how positions become fixed, and how differences intensify. Simplification, repetition, and alignment do not only influence what is communicated—they shape what is perceived as real.

Within this context, the development of understanding is not only a matter of acquiring knowledge, but of recognizing how knowledge itself is formed, interpreted, and sustained.

This shifts attention to the role of education in a broader sense.

These dynamics do not operate at a distance.

They shape how individuals perceive, interpret, and respond to the world. The formation of meaning is therefore not only a collective process, but one that returns to the individual—who must navigate within it.

Not as the transmission of established answers, but as the development of the capacity to engage with complexity—while remaining aware of the processes that shape it.

If communication can reinforce simplification, education must create space for reflection.

If perspectives tend to stabilize, education must allow for reconsideration.

If meaning is shaped within systems, education must make those systems visible.

The question is not how to eliminate difference, but how to engage with it without reducing it prematurely.

Chapter 9

Education and Thinking Without Absolutism

IF UNDERSTANDING IS SHAPED by cognitive limits, reinforced by psychological needs, structured through language, and influenced by social environments, then the question arises how these conditions can be engaged more consciously.

This question emerges as particularly relevant in the context of education.

Education is often understood as the transmission of knowledge. Information is organized, presented, and passed on with the intention of enabling individuals to understand the world more accurately. Within this model, clarity is a primary goal. Concepts are defined, distinctions are established, and answers are provided in forms that can be learned and applied.

This approach is effective in many domains. Without structured knowledge, it would be difficult to navigate complex fields such as science, mathematics, or language. Yet when education is oriented primarily toward clarity and correctness, a subtle shift can occur.

Understanding becomes associated with having the correct answer. This association is subtle but influential. When correctness develops into the primary indicator of understanding, the process through which that understanding is formed comes to be less visible. Attention shifts toward outcomes rather than conditions.

This association can be observed in familiar situations. Questions are often framed in ways that suggest a correct response, and evaluation is structured around arriving at that response efficiently. Less visible, however, is how the answer is constructed—what assumptions it depends on, what alternatives are excluded, and under what conditions it holds.

In this sense, education, like language and knowledge, is not neutral; it reflects the cultural context and dominant ideas within which it operates, shaping not only what is learned, but how reality itself is interpreted.

Over time, this can lead to a form of learning in which certainty is recognized more easily than inquiry. The ability to reproduce established answers becomes more prominent than the ability to examine how those answers are formed.

In such a framework, ambiguity is often treated as something to be resolved rather than something to be explored. Questions are valued insofar as they lead to answers, and answers are valued insofar as they can be stabilized. This creates a tendency to present knowledge as more complete than it is.

The complexity underlying that knowledge becomes less visible. What is presented as stable knowledge often rests on layers of interpretation, selection, and simplification. When these layers are not made explicit, knowledge can appear more complete than it is.

It doesn't necessarily reflect an intention to simplify excessively. It often emerges from practical constraints. Teaching requires structure. Learning requires orientation. Without some degree of simplification, knowledge would be difficult to transmit.

Yet this necessary simplification carries a risk. The risk is not that knowledge becomes incorrect, but that its provisional nature becomes obscured. What is intended as a model may be received as a final representation.

If the provisional nature of knowledge is not emphasized, what is taught as a model may be understood as a complete representation. The distinction between explanation and reality becomes less visible. Over time, this can reinforce the same movement observed in other domains: partial understanding extended toward certainty.

To think without absolutism, therefore, requires a different emphasis.

It does not mean abandoning structure or refusing to teach established knowledge. Rather, it involves maintaining an awareness of the limits of that knowledge while engaging with it. This awareness does not reduce the usefulness of knowledge. It changes how it is held. Instead of functioning as a fixed endpoint, knowledge emerges as part of an ongoing process of refinement. Education, in this sense, is not only about acquiring information, but about understanding how that information relates to what it seeks to describe.

This requires a shift in focus.

Instead of asking only *what is known*, education can also ask *how it is known*, *under what conditions it holds*, and *where its limits lie*. Such questions do not undermine knowledge; they situate it. To situate knowledge is to recognize the context in which it applies, the assumptions it depends on, and the limits beyond which it comes to be less certain. This does not weaken understanding—it makes it more precise.

This shift also changes how uncertainty is approached.

Rather than being treated as a deficiency, uncertainty points to an inherent part of complex domains. To encounter uncertainty is not necessarily to lack knowledge, but to recognize the limits of what can be known at a given moment. In this sense, the ability to remain within uncertainty emerges as a form of competence rather than a failure. It may instead indicate that one has reached the boundary of what can be clearly defined within a given framework. At such boundaries, understanding does not disappear, but changes form.

This shift appears in domains where clear answers are not readily available. In such contexts, individuals may encounter multiple explanations that each capture part of a situation without fully resolving it. The challenge is not to select a single definitive answer, but to understand how different perspectives relate to one another.

This requires a different form of engagement—one that does not move immediately toward resolution, but allows for comparison, reflection, and the recognition of limits. It involves holding multiple possibilities in view without collapsing them prematurely.

This is not an intuitive position.

As discussed in earlier chapters, there is a natural tendency to reduce uncertainty and seek stability. Education that engages with this tendency must therefore do more than present information. It must also develop to reflect on how understanding is formed and how it can become fixed. This reflective capacity is not automatic. It requires practice, exposure to different perspectives, and the willingness to reconsider what appears self-evident.

This requires cultivating a form of reflective awareness.

Such awareness includes recognizing when concepts are being treated as complete, when language is stabilizing interpretation too rigidly, and when social alignment is influencing judgment. It involves the ability to step back from a framework without immediately abandoning it.

This is a subtle but important distinction.

To think without absolutism is not to reject frameworks, but to relate to them differently. Frameworks remain necessary: they provide continuity and allow knowledge to accumulate. Without them, each situation would require complete reconsideration. Their necessity, however, does not imply completeness. They provide orientation and enable action. But they are held with an understanding that they are partial.

This has implications not only for how individuals think, but for how learning environments are structured.

In contexts where only correct answers are emphasized, there is a tendency toward convergence. Students learn to align with established frameworks, often without examining their underlying assumptions. In contrast, environments that allow for exploration, questioning, and the consideration of multiple perspectives can maintain a broader range of understanding.

Such environments do not eliminate disagreement.

But they change its function.

Disagreement emerges as a means of exploring differences in interpretation rather than a signal of error. The goal shifts from resolving differences as quickly as possible to understanding how they arise. When approached in this way, disagreement can reveal underlying assumptions that might otherwise remain implicit. It becomes a way of examining how perspectives are structured.

This approach requires time and attention. It is less efficient than simply transmitting fixed answers. Yet it allows for a deeper engagement with the material.

It also reflects the condition described throughout this book.

If no single framework fully captures reality, then the ability to move between frameworks, to examine their assumptions, and to recognize their limits emerges as essential. Education, in this sense, is not only preparation for action within a given system, but preparation for navigating multiple systems without reducing them prematurely.

This extends beyond formal education.

Education does not take place in a single domain. It unfolds across multiple contexts that interact with one another.

At the individual level, it concerns how understanding is formed—how information is interpreted, how assumptions are recognized, and how uncertainty is approached.

At the institutional level, education is structured through curricula, methods, and systems that organize knowledge and define what is to be learned. These structures provide orientation, but they also influence which perspectives are emphasized and how they are presented.

At the same time, learning takes place within social environments—family, peer groups, and broader cultural contexts. Within these settings, ideas are reinforced, challenged, or reshaped through interaction. What is accepted, questioned, or dismissed is often influenced by these relational dynamics.

These dimensions do not operate independently—they interact. Understanding is formed through their combined influence.

A further development introduces an additional dimension to these processes.

Technological systems capable of generating and organizing information are increasingly present within learning environments. They allow individuals to access knowledge rapidly, to explore questions interactively, and to connect ideas across domains in ways that were previously more difficult to achieve.

This changes the conditions under which learning takes place.

Information is no longer primarily encountered in fixed sequences, but can be approached dynamically. Questions can be refined, perspectives compared, and explanations revisited in an ongoing process. The availability of knowledge is less constrained by access, and more shaped by how it is engaged.

However, these systems operate on patterns derived from existing information. They can reveal connections, but they also reflect the structures and assumptions present in the material on which they are based.

It follows both an expansion and a limitation.

The capacity to explore increases, but so does the need to remain aware of how information is structured, selected, and presented. The challenge is not only to access knowledge more efficiently, but to engage with it more consciously.

What comes to be available expands, but the responsibility for how it is understood remains.

This shift also affects where and how meaning is formed.

Traditionally, understanding develops through interaction with others—within dialogue, shared contexts, and social environments where perspectives are exchanged, challenged, and refined. Meaning, in this sense, is shaped collectively.

With the increasing use of technological systems (AI, LLM), part of this process becomes more individualized. Interaction with information can take place directly, without immediate engagement with others. Questions can be explored, refined, and revisited within a more private and continuous process.

And yet, this does not isolate understanding from collective influence. The systems through which this interaction occurs are themselves shaped by large bodies of human-generated knowledge. What appears as an individual process remains connected to broader patterns of meaning.

This creates a shift in emphasis.

The formation of meaning may become less dependent on immediate social interaction, while remaining embedded in structures that reflect collective input. The balance between individual exploration and shared interpretation is not removed, but reconfigured.

To engage with understanding consciously therefore involves recognizing not only what is learned, but where and how it is shaped.

The capacity to think without absolutism applies to any context in which understanding is required. It influences how individuals interpret information, how they engage with others, and how they make decisions in situations that do not admit simple answers.

At its core, this capacity involves maintaining a balance.

On one side, there is the need for structure. Without it, understanding becomes diffuse and action becomes difficult. On the other side, there is the need for openness. Without it, understanding develops into rigid and unresponsive to new information.

To sustain this balance is not straightforward.

It requires ongoing engagement. It cannot be reduced to a fixed method or a set of rules. It involves continuously adjusting how one relates to knowledge, recognizing when clarity develops into simplification, and when openness develops into indeterminacy.

It doesn't lead to a final solution.

But it changes the nature of the problem.

There is no immediate solution. It requires time, attention, and a willingness to remain within questions that do not resolve quickly. Yet it allows understanding to develop without being prematurely closed.

The question is no longer how to eliminate uncertainty, but how to work within it without collapsing into either absolutism or relativism.

If education shapes how knowledge is approached, its implications do not remain within institutions. They extend into how individuals think, decide, and position themselves within a complex and changing world.

Chapter 10

The Individual in a Complex World

If the previous chapters have shown how systems of understanding are formed, stabilized, and extended toward claims of completeness, the final question is no longer theoretical.

It becomes practical.

How is one to live within this condition?

Not outside systems, but within them.

Not with complete certainty, but without abandoning understanding.

Not by rejecting structure, but without turning it into something absolute.

This question cannot be answered at the level of systems alone.

It returns to the individual.

At times, this is not only a question of interpretation, but of experience.

Moments in which the individual becomes aware of something that seems to go beyond immediate understanding.

What is encountered in such moments?

Is it a projection of the mind, or an encounter with something beyond it?

Can such an experience point toward truth, or does it only reflect the structures through which we perceive?

If it feels real—what does that imply about reality itself?

The great questions remain. For the individual in his mystic experience.

God—

What and Who is he or it?

Is he the creator of the universe, but moved on like the clockmaker?

Did he create humans as his children and are we all kind of sons of God?

How do we relate to God? As children to a father?

Does God interfere in our lives, the lives of everyone alive and dead?

Will he observe our prayers?

Can we as humans know him?

How does he communicate with us? Is that via Angels?

Is there one God or multiple Gods?

Is there a Devil who performs under the jurisdiction of that same God?

Is that to challenge mankind?

Is the Devil an autonomous entity or a creation of God?

Can we experience God through our consciousness?

If there is a God who is all loving, why is there suffering and evil in his world he creates?

Different traditions offer different interpretations. Some describe a transcendent creator, others a personal God who relates to individuals, while others understand the divine as immanent within the world itself. These perspectives shape how questions about meaning, suffering, and existence are approached, but they do not resolve them.

What is encountered in such moments?

Is it a projection of the mind, or an encounter with something beyond it?

Can such an experience point toward truth, or does it only reflect the structures through which we perceive?

And if it feels real—what does that mean?"

As an individual we are confronted with many perspectives on truth in the religious realm in trying to find answers when we have our individual mystic experience.

Living Within Frameworks Without Being Defined by Them

Human life unfolds within frameworks.

From early on, individuals are shaped by systems of meaning—cultural, linguistic, educational, and social. These frameworks provide orientation. They make the world intelligible and allow action to take place without constant reconsideration of every assumption.

Without such frameworks, life would become difficult to navigate.

And yet, as this book has suggested, these same frameworks are not complete. They represent structured ways of engaging with reality, but they do not fully capture it. To live entirely within them, without awareness of their limits, is to risk confusing the framework with what it describes.

The alternative is not to abandon frameworks.

It is to relate to them differently.

To recognize that they are necessary, but partial.

To use them, but not be fully defined by them.

To engage with them, while remaining aware that they do not exhaust reality.

This introduces a form of distance—not a rejection, but a distinction.

The Inner Dimension: Bringing Forth What Is Within

This distinction between framework and reality is not only external.

It also has an internal dimension.

The tension between outward structure and inward experience has long been recognized. In the Gospel of Thomas (Logion 70), one finds the statement:

"If you bring forth what is within you, what you bring forth will save you.

If you do not bring forth what is within you, what you do not bring forth will destroy you."[1]

The saying of Thomas does not refer to a rejection of external systems, but to the recognition that meaning cannot be fully imposed from outside. What is internal—reflection, awareness, and the capacity to engage with experience—plays an essential role.

A similar emphasis appears in the thought of Søren Kierkegaard, who argued that truth is not only something to be understood, but something to be lived. For him, the individual stands in a position where decisions must be made without complete certainty. The responsibility cannot be transferred entirely to systems, doctrines, or rational structures.

The notion of what is "within" can be misunderstood if interpreted too narrowly.

It does not refer simply to personal preference, self-expression, or the pursuit of satisfaction as it is often framed in contemporary contexts. In many settings, the idea of the self comes to be associated with achievement, fulfillment, or the optimization of experience.

In the present context, however, the self is understood differently.

It refers to the capacity to relate to one's own understanding—to recognize how perceptions, interpretations, and assumptions arise, and how they shape engagement with the world. It involves a form of internal consistency, not in the sense of fixed

1. Pagels, *Beyond Belief*, 53.

identity, but in the alignment between what is experienced, what is understood, and how one responds.

This does not imply a fixed state of balance.

Rather, it involves an ongoing process of adjustment, in which one remains attentive to both internal experience and external conditions. It includes the ability to question one's own interpretations, to recognize when certainty is premature, and to remain open without losing orientation.

To exist is to choose.

And to choose is to accept the uncertainty that comes with it.

So, it fails to eliminate anxiety completely. As discussed earlier, the openness of possibility introduces a form of tension that cannot be fully resolved. Yet this tension is not only a burden—it is also a condition for insight.

Without it, there would be no need to reflect.

Holding Understanding Without Closing It

To live within this condition requires a particular way of relating to understanding.

This condition is not new—it reflects a long-standing structure of human experience.

Human beings have lived within uncertainty long before the development of formal systems of thought. For most of human history—long before written traditions and structured knowledge—life unfolded in small communities where survival depended on rapid interpretation, trust within the group, and the ability to respond to potential threat.

In such environments, uncertainty was not an abstract concept. It was immediate and consequential. The inability to interpret a situation quickly could have direct consequences for survival. As a result, there was a strong tendency to reduce complexity into more immediate and actionable forms. What was unclear had to become clear, even if that clarity was incomplete.

This tendency has not disappeared.

Although the conditions of life have changed, the underlying structures remain. Human cognition still operates in ways that prioritize orientation over completeness. Faced with uncertainty, there is a natural movement toward meaning, toward explanation, and toward positions that provide stability.

This movement is not in itself problematic. It allows action, coordination, and continuity. Without it, decision-making would become difficult, and engagement with the world would lose direction.

This movement toward orientation can be understood more clearly when considering the role of the individual.

For Jean-Paul Sartre [2], human existence is defined not only by the need to interpret the world, but by the necessity to act within it. Choices are not made from a position of complete certainty. They are made within conditions that are always partially understood.

To act, therefore, is to take responsibility without full knowledge of outcomes.

This responsibility is not limited to the individual in isolation. It unfolds within a social context. Decisions contribute to the conditions within which others also act, and are themselves shaped by those conditions.

In this sense, the self is not fixed or given. It is formed through engagement—through the way one responds to situations, to others, and to the limits of one's own understanding.

Yet the same movement that enables action also carries a risk.

When complexity is reduced too quickly, what is gained in clarity may be lost in depth. Interpretations that are useful in a practical sense may not fully correspond to the reality they seek to describe. What begins as a necessary simplification can become a fixed framework, within which alternative possibilities are no longer considered.

This pattern can be observed across different contexts.

In situations of uncertainty, individuals and groups often move toward clearer distinctions, more defined interpretations, and stronger forms of alignment. What is ambiguous becomes

2. Sartre, *Existentialism Is a Humanism.*

categorized. What is complex becomes simplified. What is uncertain becomes framed in ways that allow for immediate response.

These processes are effective. But they can also lead to the premature closure of understanding.

To recognize this is not to reject the need for orientation. It is to become aware of the conditions under which orientation takes shape, and of the point at which it begins to limit rather than support understanding.

It involves holding beliefs without turning them into absolutes.

This does not mean that all beliefs are equal, or that no judgment can be made. As explored throughout this book, some interpretations are more coherent, more consistent, and more responsive to evidence than others. Distinctions remain necessary.

But those distinctions are not final.

They are held with an awareness that they may be revised.

This creates a position that is neither absolutist nor relativist.

Not absolutist, because it does not claim complete certainty.

Not relativist, because it does not abandon the possibility of better or worse understanding.

Instead, it maintains a form of engagement that remains open.

This openness is not passive. It requires effort. It involves recognizing when simplification takes shape as reduction, when language becomes fixed, and when social alignment begins to shape judgment more than reflection.

It requires the capacity to reconsider.

The Individual Within Society

The individual does not exist outside social structures.

As discussed in earlier chapters, communication, identity, and group dynamics influence how perspectives are formed and maintained. To think independently is not to think in isolation, but to remain aware of these influences while engaging with them.

This awareness does not remove the pressures of alignment or belonging.

But it changes how they are experienced.

Instead of fully identifying with a perspective, one can recognize it as a position within a broader field of possibilities. Instead of reducing opposing views to error, one can examine the processes through which they arise.

This is not to say that it does eliminate disagreement.

But it transforms its role.

Disagreement comes to be an opportunity for understanding rather than a threat to stability. In practice, this shift can be observed in various forms of collaboration that bring together different perspectives without requiring immediate agreement. In scientific research, for example, progress often depends on the ability to work across differing hypotheses and interpretations, allowing competing ideas to be tested rather than resolved prematurely.

Similar patterns appear in other domains. International cooperation in areas such as space exploration or global health requires coordination between groups that do not share identical frameworks yet are able to work toward common objectives. These efforts do not eliminate differences but operate alongside them.

What becomes visible in such contexts is not the absence of disagreement, but the capacity to engage with it without reducing it to opposition.

When Understanding Is Disrupted

The condition described throughout this book is not static.

It is subject to change.

Frameworks of understanding depend not only on their internal structure, but on the limits of what is currently known. When those limits shift, the frameworks built upon them may be challenged in ways that are not easily absorbed.

Some changes are gradual.

Others are more disruptive.

The discovery of life beyond Earth, for example, would not only extend biological knowledge. It would challenge assumptions about the uniqueness of human existence. Systems that place humanity at the center of meaning would be required to adapt.

The emergence of new historical sources—texts that present alternative accounts of foundational figures—could introduce tensions within established traditions. The question would not only be whether such sources are accurate, but how they can be integrated into existing narratives without altering their structure.

Advances in the study of consciousness may further complicate the understanding of the self. If subjective experience is shown to arise in ways that challenge current assumptions, or if forms of awareness are identified beyond the human domain, the distinction between subject and object may become less stable.

Developments in artificial intelligence introduce similar questions. If systems are capable of forms of reasoning or behavior that resemble understanding, the position of the human mind as uniquely central may be reconsidered.

In cosmology, the expansion of knowledge—whether through the study of the primordial gravitational waves that may tell us what happened in the earliest stage after the Big Bang—continues to extend the scale of reality. What once appeared comprehensive develops into partial within a larger context.

These examples do not point to a single conclusion.

At the same time, responses to such shifts are not only reactive. Across different domains, there are ongoing efforts to develop ways of engaging with increasing complexity. Interdisciplinary approaches, open scientific collaboration, and new forms of dialogue seek to bring together perspectives that were previously separated.

These efforts do not resolve uncertainty.

But they reflect an awareness that understanding cannot be contained within a single framework. They represent attempts to remain responsive to change while maintaining continuity.

They illustrate a pattern.

Understanding is always situated within the limits of what is known. When those limits shift, the structures built within them are exposed to revision.

This is no way necessarily led to collapse.

But it reveals that stability is conditional.

Remaining Within the Question

To live within this condition is not to seek final resolution.

It is to remain within the question.

Not as a problem to be solved once and for all, but as a context in which understanding takes place.

The question of existence—the one with which this book began—does not disappear. It remains present, even as systems attempt to answer it. It is not replaced by explanation but accompanied by it.

To remain within the question is not to abandon knowledge.

It is to recognize its limits without rejecting its value.

Final Position

Human perception does not stand outside reality, observing it from a position of complete clarity.

We are within it.

We construct frameworks to understand it.

We use language to express it.

We rely on systems to navigate it.

And yet, none of these fully contain it.

To live without absolutism is not to live without direction.

It is to act with awareness of limitation.

To engage without claiming finality.

To hold understanding without closing it.

This stance is not simple.

It does not provide the certainty of fixed systems, nor the ease of indifference.

But it remains responsive.

Different approaches to understanding reflect this tension in distinct ways.

Some thinkers emphasize the need for modesty—an awareness of limits, openness to multiple perspectives, and a reluctance to reduce complexity prematurely. Others take a more assertive role, advancing bold interpretations that challenge existing frameworks and push understanding forward.

Both movements have their function.

Without the first, understanding risks becoming rigid and exclusionary. Without the second, it may lack the impetus to move beyond established positions.

The development of knowledge appears to depend not on one of these approaches alone, but on their interaction.

What appears at one moment as a necessary disruption may, over time, become a new point of stability.

Final Reflection

Technological developments continue to change the conditions under which understanding takes place. Systems capable of generating and connecting information make it possible to explore questions in ways that were not previously available. They can assist in identifying patterns, comparing perspectives, and extending lines of inquiry.

Yet these developments do not replace the fundamental conditions of human experience.

The tendency to ask questions, to seek meaning, and to wonder about what lies beyond current understanding remains unchanged. These are not properties of the systems we create, but of the beings who create and engage with them.

What may change is how we explore.

What remains is that we continue to ask.

Whatever meaning emerges is not something we fully discover or define, but something that takes shape in how we engage—with others, with the world, and with the limits of our own understanding.

Not in certainty.

But in awareness.

In this sense, the question is not only what we understand today, but what may still emerge—and how new answers might reshape the way we understand reality.

We do not resolve existence. We live within it—while continually shaping the way we understand it.

If understanding is shaped by shifting knowledge, competing perspectives, and evolving conditions, the question becomes how one can move within it.

Not by resolving uncertainty, but by learning to engage with it.

Chapter 11

Navigating Competing Truths

If the previous chapters have shown how understanding is formed, stabilized, and extended toward claims of completeness, the final question is not only how these processes occur, but how they are lived—not in theory, but in practice.

Human beings do not experience reality as a neutral field. They encounter it through perception, language, systems of meaning, and histories that shape interpretation. What appears as self-evident is often the result of these layers.

To navigate within this condition is not to step outside it.

It is to move within it with awareness.

The Search for Orientation

The need to find meaning and direction is not recent. It is rooted in conditions that extend far beyond the development of formal knowledge.

For most of human history—long before written traditions and structured systems—life unfolded in small communities where survival depended on rapid interpretation, trust within the group, and sensitivity to potential threat. What was unfamiliar could not always be examined in detail. It had to be categorized quickly—as safe or unsafe, as belonging or not belonging.

This tendency remains.

Even in contexts where survival is no longer immediately at stake, the structure of response persists. What is unfamiliar may still be approached with caution. What differs may be interpreted as incompatible. The distinction between "within" and "outside" continues to shape perception.

This does not arise from a single cause.

It reflects a recurring pattern.

From Survival to Structure

Over time, these patterns become embedded in systems of belief.

Shared interpretations provide stability. They define what is considered true, what is valued, and how individuals relate to one another. These systems do not emerge arbitrarily. They develop within specific historical and cultural conditions.

Yet they are often experienced as given.

What is constructed over time appears as self-evident. What has developed through interaction develops into fixed as structure. In this way, systems of meaning extend beyond their initial function and take on a more comprehensive role.

They begin to define reality.

The Risk of Simplification

In situations of uncertainty, there is a natural movement toward clarity.

Complexity is reduced into more defined forms. Ambiguity is resolved into distinction. What is difficult to interpret becomes categorized in ways that allow for action.

This process is necessary.

Yet it introduces a risk.

When simplification becomes fixed, new insights and alternative perspectives become easier to exclude.

The result is not only misunderstanding but narrowing.

Reflection and Position

To navigate without absolutism does not mean abandoning belief.

It means relating to belief differently.

Every position is shaped by conditions—by background, experience, culture, and biological predispositions that influence perception and response. What appears as a neutral judgment is often situated within these layers.

To reflect on this is not to invalidate understanding.

It is to recognize its context.

Such recognition introduces a form of distance—

not a rejection of belief, but an awareness that belief is not identical to reality.

The formation of meaning does not occur in isolation.

It is shaped by conditions that extend beyond individual reflection. Family context, cultural background, and the broader social environment influence how perspectives are encountered and interpreted. What is considered plausible, acceptable, or meaningful is often embedded within these structures before it emerges as a matter of conscious choice.

In contemporary contexts, additional layers influence this process.

Information is increasingly presented through frames that emphasize clarity, effectiveness, and emotional resonance. Complex situations are reduced into forms that can be quickly recognized and communicated. These reductions are not neutral. They make use of tendencies in human perception—toward familiarity, coherence, and immediate understanding.

Technological systems further amplify these dynamics.

Algorithms and digital platforms influence which information is encountered, how often it is repeated, and in what form it appears. What appears is not only a reflection of reality, but of the structures through which it is filtered and presented.

It doesn't eliminate the possibility of forming meaning.

But it changes the conditions under which it takes place.

The construction of meaning becomes less a passive reception of information, and more an active process that requires attention and effort. Without such engagement, there is a tendency to adopt perspectives that are readily available, familiar, or reinforced through repetition.

The effect of this is a further tension between expansion and simplification.

The capacity to explore meaning is expanded, yet the tendency toward simplification is intensified. What emerges as accessible increases, but so does the influence of the structures that organize it.

To navigate within this condition is not to reject these influences, but to remain aware of them.

Ways of Navigating Within Uncertainty

Different responses to uncertainty can be observed, not as fixed principles, but as recurring ways in which human beings orient themselves.

Some respond through commitment. When certainty cannot be reached, a decision is still made. Søren Kierkegaard [1] described this as a leap of faith—not the absence of reflection, but the recognition that action cannot always wait for complete understanding. What matters is not that doubt disappears, but that it does not prevent engagement.

Others turn toward what can be examined and tested. Scientific inquiry does not remove uncertainty, but it structures it. It defines what can be measured, compared, and revised. Within this domain, understanding comes to be shared and verifiable, even if it remains incomplete.

There are also forms of orientation that begin from experience itself. Consciousness is not only something that can be studied from the outside, but something that is lived from within.

1. Kierkegaard, *Fear and Trembling.*

Perception, awareness, and reflection provide a dimension of understanding that cannot be fully reduced to external observation.

At the same time, individuals continue to rely on broader frameworks of meaning. Religious and philosophical traditions offer structures through which existence can be interpreted. Yet these structures are often expressed through texts and formulations that reflect specific historical contexts. To engage with them does not require treating those expressions as final. It allows for the possibility that what they point toward exceeds their literal form—that meaning is not exhausted by the words through which it is conveyed.

The result of this is that it creates the possibility that what is referred to as "God", or ultimate reality, is not fully contained within any single formulation. What is expressed in tradition may be understood as an approach, rather than a complete representation.

In another direction, thinkers such as Daniel Dennett [2] have argued that human responsibility can be maintained even within a structured and causal world. From this perspective, the presence of underlying processes does not eliminate the meaningful distinction between actions, decisions, and accountability. Rather than requiring absolute freedom, responsibility can be understood as emerging within the conditions that shape behavior.

These approaches do not converge into a single system.

Yet each reflects an attempt to respond to the same condition: how to act, interpret, and relate within a reality that is not fully accessible.

To move between them is not to lose direction.

It is to recognize that orientation itself can take different forms.

2. Dennett, *Freedom Evolves.*

Future Horizons

Understanding is not static.

Developments in science, technology, and historical research continue to expand the limits of what is known. Discoveries related to consciousness, artificial intelligence, or the structure of the universe may introduce perspectives that challenge current frameworks.

What appears stable today may be reinterpreted tomorrow.

This not really undermine understanding.

It reveals its provisional nature.

Current developments already indicate how such shifts may continue to unfold.

In cosmology, increasingly refined observations—such as those based on gravitational wave detection and deep-space imaging—extend the ability to examine processes that were previously beyond reach. While the general outline of cosmic development is broadly shared, questions remain about its earliest stages and underlying structure.

In neuroscience, advances in imaging and modeling continue to deepen the understanding of how the brain processes information. Yet the relation between neural activity and subjective experience remains unresolved, leaving space for further insight into the nature of consciousness.

In archaeology and the study of early human societies, new discoveries continue to revise earlier interpretations. Patterns once considered stable are reconsidered as additional evidence emerges, suggesting that historical development may be more complex and varied than previously assumed.

These developments do not point toward a single transformation.

They reflect a recurring pattern: what is established expands at its boundaries. What appears complete reveals new questions. What is understood becomes part of a larger context.

It doesn't diminish what is known.

It suggests that understanding remains open—not only limited, but capable of further development.

Living Within the Question

To navigate within this condition is not to resolve uncertainty, but to remain within it without collapsing into either absolutism or indifference.

It involves engaging with knowledge without treating it as final; recognizing limits without abandoning meaning and acting within frameworks while remaining aware of their scope.

This position does not provide immediate clarity.

It requires time, attention, and the willingness to remain within questions that do not resolve quickly.

What remains is the persistence of the question itself.

What changes are the ways in which we explore.

What remains is the persistence of the question itself.

Yet this persistence is not experienced in isolation.

Even as perspectives differ and understanding remains open, there are conditions that are shared.

Why there is something rather than nothing remains one of the fundamental questions of human existence.

How to live within that condition may be an even more immediate one.

Human beings navigate between conscious experience, inner intuition, and the structures of society within which meaning is shaped. None of these provide complete certainty. Yet within them, individuals continue to search for orientation, understanding, and ways of living that move toward wisdom rather than closure.

Perhaps this is not a condition to overcome, but one to navigate.

Like wayfarers searching for direction rather than final arrival.

Chapter 12

Common Ground

If the ways in which we explore continue to change, the question becomes not only how perspectives differ, but what remains accessible across them.

What We Still Share

If the previous chapters have shown how perspectives form, stabilize, and diverge, a further question remains. Not how differences arise, but what persists despite them.

Human understanding is not only divided. It is also shared.

Across cultures, disciplines, and traditions, there are points of convergence that do not eliminate difference but provide a basis from which it can be engaged. These points do not form a complete system. They do not resolve disagreement. But they indicate that, beneath competing interpretations, there remains a common ground.

A Shared World

Despite differences in interpretation, human beings do not inhabit separate realities. They live within a shared physical world governed by structures that can be examined, tested, and, to a significant extent, agreed upon.

Over the past century, a broad scientific picture has emerged that, while incomplete, is widely accepted across disciplines.

The Earth is understood to have formed approximately 4.5 billion years ago as part of the development of the solar system. The Sun, planets, and other bodies emerged from a rotating cloud of gas and dust. The Moon likely formed from a massive collision early in Earth's history. These processes are supported by geological and astronomical evidence.

At a larger scale, the universe itself has evolved over vast timescales. Observations of cosmic background radiation suggest an early, dense state from which the universe expanded. Galaxies formed, stars ignited, and heavier elements were created through stellar processes.

On Earth, life emerged and developed through evolution. All known living organisms share a common genetic structure based on DNA, indicating a shared origin. Over time, natural selection, mutation, and environmental pressures led to the diversity of life we observe today.

Geological and biological processes are closely linked. Plate tectonics has continuously reshaped continents. Volcanic activity contributed to atmospheric development. Photosynthesis gradually increased oxygen levels, enabling more complex life forms to emerge. Periods of mass extinction reset ecosystems, allowing new forms of life to develop.

For readers less familiar with these domains, the key point is not every detail, but the level of agreement: across cultures and scientific communities, these broad patterns are widely shared. They form part of what can reasonably be called shared knowledge.

Yet this shared understanding is not static. It continues to evolve.

In cosmology, new instruments extend what can be observed. Space-based telescopes are now able to look further back in time, raising questions about how early structures formed and whether current models fully explain them. Dark matter and dark energy—thought to make up most of the universe—remain only partially understood.

The discovery of planets beyond our solar system has expanded the scope of inquiry. Thousands of exoplanets have been identified, and ongoing research explores whether conditions for life—such as water and stable atmospheres—may exist elsewhere.

In neuroscience, significant progress has been made in understanding how the brain processes information, regulates behavior, and constructs perception. However, the relationship between physical brain activity and subjective experience—consciousness—remains one of the most open questions in science.

These developments illustrate a key point: even where there is strong agreement, there are also frontiers. What is shared is not only a body of knowledge, but an ongoing process of inquiry.

A Shared Human Condition

Beyond the physical world, there is also a shared condition of existence.

Across different cultures and historical periods, human beings encounter similar fundamental situations. They are born into conditions they did not choose. They depend on others. They experience loss, uncertainty, and limitation. They are aware of their own mortality.

Meanwhile, they seek meaning, connection, and orientation. They ask questions about purpose, value, and truth. These questions are not only philosophical. They appear in everyday decisions—how to respond to uncertainty, how to relate to others, and how to act when outcomes are not fully known. These questions take different forms, but they recur.

This shared condition does not produce identical answers. It produces a field within which answers are sought.

Between Defense and Exploration

Human behavior reflects more than one inheritance.

On one side, there is a tendency toward protection and preservation. For most of human history, survival depended on maintaining group cohesion and responding cautiously to what was unfamiliar. This has left traces in how difference is perceived and how boundaries are formed.

On the other side, there is a capacity for exploration. Human beings do not only preserve what exists; they also question, experiment, and extend beyond established limits. They form connections across groups, exchange ideas, and develop new ways of understanding.

These tendencies do not exclude one another—they coexist.

Civilizations develop not by eliminating one in favor of the other, but by negotiating their relationship. Preservation provides stability. Exploration introduces change. When either dominates without balance, understanding becomes either rigid or unstable.

Shared Capacities

What allows movement beyond fixed positions is not agreement, but capacity.

Human beings possess the ability to reflect on their own understanding. They can examine assumptions, reconsider interpretations, and revise beliefs considering new information. They are capable of cooperation beyond immediate group boundaries, of learning across traditions, and of engaging with perspectives that differ from their own.

Among these capacities, one stands out as particularly fundamental: curiosity.

Curiosity does not guarantee correct understanding. It does not eliminate error. But it creates the possibility of movement. It allows individuals and communities to move beyond what is given, to question what appears fixed, and to explore what is not yet understood.

Throughout history, this capacity has played a decisive role. It has led to the development of knowledge, the transformation of societies, and the expansion of what is considered possible.

Convergences Without Uniformity

Within different traditions, certain orientations appear repeatedly. For readers less familiar with religious traditions, it is helpful to first identify their central concerns.

The major monotheistic traditions—Judaism, Christianity, and Islam—along with earlier influences such as Zoroastrianism, are built around a few core ideas.

They affirm that reality has a single ultimate source. They describe a relationship between human beings and this source, often mediated through revelation, prophets, or sacred texts. Human actions are understood to have moral significance, and life is framed within a broader horizon of accountability, justice, and meaning.

These traditions also emphasize ethical life: care for others, responsibility, and the importance of intention. While doctrines differ, the idea that how one lives matters is central.

In Christianity, this is often expressed through love and grace. In Judaism, through covenant and law. In Islam, through submission to the will of God and ethical conduct. Zoroastrianism places strong emphasis on the struggle between good and evil and the role of human choice within it.

These shared elements are often described as ethical monotheism, where belief in one ultimate reality is closely linked to how one ought to live.

Eastern traditions approach similar questions differently.

In traditions such as Buddhism and Hindu philosophical schools, the focus often shifts from an external source of authority to the nature of experience itself. Questions center on suffering, attachment, awareness, and the possibility of liberation or balance.

Practices such as meditation, reflection, and disciplined living aim to transform how reality is experienced rather than to define it in fixed doctrinal terms. Yet the underlying concern remains comparable: how to live, how to understand existence, and how to relate to what is beyond immediate control.

A related convergence can be found in Greek philosophy.

Thinkers such as Plato [1] and Aristotle [2] did not rely on revelation, but they addressed similar questions. Plato points toward a higher order of reality and the idea of the Good. Aristotle emphasizes reason, purpose, and virtue as ways of living well.

Later traditions such as Stoicism [3] develop ideas of inner discipline, acceptance of what cannot be controlled, and alignment with a rational order of the universe.

What unites these approaches is not agreement in doctrine, but a shared orientation.

Human life gives rise to fundamental questions. Ethical conduct is understood to matter. Reality may extend beyond immediate perception, and understanding is seen as something that requires effort, reflection, and discipline.

What differs is how these insights are expressed—through revelation, philosophy, or practice.

This suggests that what is common is not a single belief, but a recurring attempt to respond to the same human condition.

Building Without Finality

If common ground exists, it does not provide a final foundation.

It does not resolve all differences or establish a single framework that replaces others. Instead, it offers a starting point—a basis from which further understanding can develop without requiring complete agreement.

To build on such a ground is not to eliminate tension. It is to work within it.

1. Plato, *Republic.*
2. Aristotle, *Nicomachean Ethics.*
3. Epictetus, *Enchiridion.*

This implies recognizing what can be shared without assuming that everything must be unified. It involves engaging with difference without immediately reducing it to error or opposition.

A Continuing Process

Understanding does not reach a final state.

What is shared today may be revised tomorrow. New discoveries may extend or alter what is currently accepted. New interpretations may reshape how existing knowledge is understood.

Common ground, therefore, is not fixed.

It is part of an ongoing process in which knowledge develops, perspectives shift, and new connections emerge.

Final Reflection

Human beings do not begin from nothing.

They inherit a world that is already structured, a history that has already unfolded, and capacities that allow them to engage with both.

They share more than is often recognized.

Not a single truth that resolves all questions, but a set of conditions within which understanding becomes possible.

To recognize this is not to diminish difference.

It is to see that, even within divergence, there remains something that can be built upon.

Not as a final answer, but as a starting point from which further understanding can continue to develop.

Perhaps, as already suggested in the ancient wisdom associated with Zarathustra [4], the task is not to claim possession of final truth, but to continue seeking a more truthful orientation toward reality.

4. Boyce, *Zoroastrians.*

To develop greater awareness, honesty, and responsibility; to approach others through constructive communication; and to turn insight into actions that contribute not only to oneself, but also to others.

Not toward a final state of perfection, but toward greater wisdom, harmony, and depth in human life.

Bibliography

Afshordi, Niayesh, and Phil Halper. *Battle of the Big Bang: The New Tales of Our Cosmic Origins*. Chicago: University of Chicago Press, 2022.

Allen, James P., trans. *The Ancient Egyptian Pyramid Texts*. Atlanta: Society of Biblical Literature, 2005.

Aquinas, Thomas. *Summa Theologica*. Translated by Fathers of the English Dominican Province. New York: Benziger Bros., 1947. Originally published 1265–1274.

Aristotle. *Nicomachean Ethics*. Translated by Roger Crisp. Cambridge: Cambridge University Press, 2000.

Armstrong, Karen. *A History of God: The 4,000-Year Quest of Judaism, Christianity, and Islam*. New York: Ballantine, 1993.

Asch, Solomon. "Opinions and Social Pressure." *Scientific American* 193, no. 5 (1955): 31–35.

Assmann, Jan. *Death and Salvation in Ancient Egypt*. Ithaca, NY: Cornell University Press, 2005.

Aurelius, Marcus. *Meditations*. Translated by Martin Hammond. London: Penguin Classics, 2006.

Bender, Emily M., Timnit Gebru, Angelina McMillan-Major, and Margaret Mitchell. "On the Dangers of Stochastic Parrots: Can Language Models Be Too Big?" In *Proceedings of the 2021 ACM Conference on Fairness, Accountability, and Transparency*, 610–23. New York: Association for Computing Machinery, 2021.

The Bible. New Revised Standard Version. Grand Rapids, MI: Zondervan, 1989.

Bostrom, Nick. *Superintelligence: Paths, Dangers, Strategies*. Oxford: Oxford University Press, 2014.

Boyce, Mary. *Zoroastrians: Their Religious Beliefs and Practices*. London: Routledge, 1979.

Boyer, Pascal. *Religion Explained: The Evolutionary Origins of Religious Thought*. New York: Basic Books, 2001.

Carroll, Sean. *The Big Picture: On the Origins of Life, Meaning, and the Universe Itself*. New York: Dutton, 2016.

Collins, John J. *The Apocalyptic Imagination: An Introduction to Jewish Apocalyptic Literature*. 3rd ed. Grand Rapids, MI: Eerdmans, 2016.

Damasio, Antonio. *Descartes' Error: Emotion, Reason, and the Human Brain*. New York: Putnam, 1994.

———, and Hanna Damasio. *Self Comes to Mind: Constructing the Conscious Brain*. New York: Pantheon, 2010.

Darwin, Charles. *On the Origin of Species*. London: John Murray, 1859.

Dawkins, Richard. *The God Delusion*. London: Bantam Press, 2006.

———. *The Selfish Gene*. Oxford: Oxford University Press, 1976.

de Waal, Frans. *The Bonobo and the Atheist: In Search of Humanism among the Primates*. New York: W. W. Norton, 2013.

Dehaene, Stanislas. *Consciousness and the Brain: Deciphering How the Brain Codes Our Thoughts*. New York: Viking, 2014.

Dennett, Daniel C. *Freedom Evolves*. New York: Viking, 2003.

Descartes, René. *Meditations on First Philosophy*. Translated by John Cottingham.

Dewey, John. *How We Think*. Mineola, NY: Dover, 1997. Originally published 1910.

Dostoevsky, Fyodor. *Crime and Punishment*. Translated by David McDuff. London: Penguin Classics, 2002.

———. *The Brothers Karamazov*. Translated by Richard Pevear and Larissa Volokhonsky. New York: Farrar, Straus and Giroux, 1990.

Easwaran, Eknath, trans. *The Upanishads*. 2nd ed. Tomales, CA: Nilgiri Press, 2007.

Ehrman, Bart D. *Jesus: Apocalyptic Prophet of the New Millennium*. Oxford: Oxford University Press, 1999.

———. *Misquoting Jesus: The Story Behind Who Changed the Bible and Why*. New York: HarperOne, 2005.

———. "The Gospel of John: Who Wrote It, When and Why?" *The Bart Ehrman Blog*, February 20, 2025. https://ehrmanblog.org/the-gospel-of-john-who-wrote-it-when-and-why/.

Epictetus. *Enchiridion*. Translated by George Long. Mineola, NY: Dover, 2004.

Festinger, Leon. *A Theory of Cognitive Dissonance*. Stanford, CA: Stanford University Press, 1957.

Freire, Paulo. *Pedagogy of the Oppressed*. 30th anniversary ed. New York: Continuum, 2000.

Freud, Sigmund. *The Ego and the Id*. New York: W. W. Norton, 1960. Originally published 1923.

Gadamer, Hans-Georg. *Truth and Method*. 2nd rev. ed. Translated by Joel Weinsheimer and Donald G. Marshall. New York: Continuum, 2004.

George, Andrew, trans. *The Epic of Gilgamesh*. London: Penguin Classics, 1999.

Graeber, David, and David Wengrow. *The Dawn of Everything: A New History of Humanity*. London: Allen Lane, 2021.

Haidt, Jonathan. *The Righteous Mind: Why Good People Are Divided by Politics and Religion*. New York: Pantheon, 2012.

Harari, Yuval Noah. *Sapiens: A Brief History of Humankind*. London: Harvill Secker, 2011.

Harris, Annaka. *Conscious: A Brief Guide to the Fundamental Mystery of the Mind*. New York: Harper, 2019.

Harris, Sam. *The End of Faith: Religion, Terror, and the Future of Reason*. New York: W. W. Norton & Company, 2004.

Hegel, G. W. F. *Phenomenology of Spirit*. Translated by A. V. Miller. Oxford: Oxford University Press, 1977.

Hoffman, Donald. *The Case Against Reality: Why Evolution Hid the Truth from Our Eyes*. New York: W. W. Norton & Company, 2019.

Holland, Tom. *Dominion: How the Christian Revolution Remade the World*. New York: Basic Books, 2019.

Hornung, Erik. *The Ancient Egyptian Books of the Afterlife*, translated by David Lorton (Ithaca, NY: Cornell University Press, 1999.

Irani, D. J. *Understanding the Gathas: The Hymns of Zarathushtra*. Edited by Kaikhosrov D. Irani. Costa Mesa, CA: Mazda, 1994.

Jung, Carl G. *The Archetypes and the Collective Unconscious*. 2nd ed. Translated by R. F. C. Hull. Princeton, NJ: Princeton University Press, 1981.

Kahneman, Daniel. *Thinking, Fast and Slow*. New York: Farrar, Straus and Giroux, 2011.

Kaizan, Khosro (Pardis). *The Gathas: The Sublime Book of Zarathustra*. Charleston, SC: Createspace Independent, 2007.

Kant, Immanuel. *Critique of Pure Reason*. Translated and edited by Paul Guyer and Allen W. Wood. Cambridge: Cambridge University Press, 1998. Originally published 1781.

Kierkegaard, Søren. *Fear and Trembling*. Translated by Alastair Hannay. London: Penguin Classics, 1985.

———. *The Concept of Anxiety*. Translated by Reidar Thomte. Princeton, NJ: Princeton University Press, 1980. Originally published 1844.

van Kooten, George H. *The Gospel of John and Hellenistic Philosophy*. Cambridge: Cambridge University Press, 2008.

———. *Reverberations of Good News: The Gospels in Context, Then and Now*. Utrecht: Kok, 2026.

Kuhn, Thomas S. *The Structure of Scientific Revolutions*. Chicago: University of Chicago Press, 1962.

Laozi. *Tao Te Ching*. Harmondsworth: Penguin Classics, 1963.

Lennox, John C. *God's Undertaker: Has Science Buried God?* Oxford: Lion, 2007.

MacGregor, Neil. *A History of the World in 100 Objects*. London: Allen Lane, 2010.

McLuhan, Marshall. *Understanding Media: The Extensions of Man*. Cambridge, MA: MIT Press, 1994. Originally published 1964.

Meyer, Marvin, ed. *The Nag Hammadi Scriptures: The Revised and Updated Translation of Sacred Gnostic Texts Complete in One Volume*. New York: HarperOne, 2007.

Meyer, Stephen C. *The Return of the God Hypothesis: Compelling Scientific Evidence for the Existence of God*. New York: HarperOne, 2021.

Nietzsche, Friedrich. *Beyond Good and Evil*. Translated by Walter Kaufmann. New York: Vintage Books, 1966.

———. *The Gay Science*. Translated by Walter Kaufmann. New York: Vintage Books, 1974.

———. *Thus Spoke Zarathustra*. Translated by Adrian Del Caro. Cambridge: Cambridge University Press, 2006.

Noble, Denis. *The Music of Life: Biology beyond Genes*. Oxford: Oxford University Press, 2006.

Pagels, Elaine. *Beyond Belief: The Secret Gospel of Thomas*. New York: Random House, 2004.

Plato. *Republic*. Translated by G. M. A. Grube. Revised by C. D. C. Reeve. Indianapolis: Hackett, 1992.

Popper, Karl. *Conjectures and Refutations: The Growth of Scientific Knowledge*. London: Routledge & Kegan Paul, 1963.

———. *The Logic of Scientific Discovery*. London: Routledge, 1959.

Postman, Neil. *Amusing Ourselves to Death: Public Discourse in the Age of Show Business*. New York: Penguin, 1985.

The Qur'an. Translated by M. A. S. Abdel Haleem. Oxford: Oxford University Press, 2004.

Rahula, Walpola. *What the Buddha Taught*. New York: Grove, 1959.

"Rig Veda 10.129 (Nasadiya Sukta)." Translated by Wendy Doniger O'Flaherty. In *The Rig Veda: An Anthology: One Hundred and Eight Hymns, Selected, Translated and Annotated*, 25–26. London: Penguin Classics, 1981.

Robinson, James M., ed. *The Nag Hammadi Library in English*. 3rd rev. ed. San Francisco: HarperSanFrancisco, 1990.

Rovelli, Carlo. *Reality Is Not What It Seems: The Journey to Quantum Gravity*. New York: Riverhead, 2017.

Sagan, Carl. *Cosmos*. New York: Random House, 1980.

Sapolsky, Robert M. *Determined: A Science of Life without Free Will*. New York: Penguin, 2023.

Sartre, Jean-Paul. *Existentialism Is a Humanism*. Paris: Nagel, 1946. English ed. New Haven, CT: Yale University Press, 2007.

Schmidt, Klaus. *Göbekli Tepe: A Stone Age Sanctuary in South-Eastern Anatolia*. Berlin: ex oriente, 2012.

Schopenhauer, Arthur. *The World as Will and Representation*. 2 vols. Translated by E. F. J. Payne. New York: Dover, 1969.

Seth, Anil. *Being You: A New Science of Consciousness*. London: Faber & Faber, 2021.

Shaw, Julia. *Making Evil: The Science Behind Humanity's Dark Side*. New York: Bantam, 2019.

Simon, Herbert A. *Models of Bounded Rationality*. 2 vols. Cambridge, MA: MIT Press, 1982.

Smith, Mark S. *The Priestly Vision of Genesis 1*. Minneapolis: Fortress, 2010.

Spinoza, Baruch. *Ethics*. Translated by Edwin Curley. London: Penguin Classics, 1996.

Swaab, Dick. *We Are Our Brains: From the Womb to Alzheimer's*. New York: Spiegel & Grau, 2014.

"The Gospel of Thomas." In *The Nag Hammadi Scriptures: The Revised and Updated Translation of Sacred Gnostic Texts Complete in One Volume*, edited by Marvin Meyer, 137–76. New York: HarperOne, 2007.

"The Gospel of Thomas," Saying 70. Translated by George MacRae. In *The Nag Hammadi Library in English*, edited by James M. Robinson, 138. San Francisco: HarperSanFrancisco, 1990.

Watzlawick, Paul, Janet H. Beavin, and Don D. Jackson. *Pragmatics of Human Communication: A Study of Interactional Patterns, Pathologies, and Paradoxes*. New York: W. W. Norton, 1967.

Wittgenstein, Ludwig. *Philosophical Investigations*. Oxford: Blackwell, 1953.

———. *Tractatus Logico-Philosophicus*. Translated by C. K. Ogden. London: Routledge, 1922.

Zuboff, Shoshana. *The Age of Surveillance Capitalism: The Fight for a Human Future at the New Frontier of Power*. New York: PublicAffairs, 2019.

www.ingramcontent.com/pod-product-compliance
Lightning Source LLC
LaVergne TN
LVHW050824061126
851857LV00012B/958